GCE O-Level F

Physics

B. P. Brindle, B.Sc.
Edited by D. F. Erskine

Published by Intercontinental Book Productions
in conjunction with Seymour Press Ltd.
Distributed by Seymour Press Ltd.,
334 Brixton Road, London, SW9 7AG

Published 1976 by Intercontinental Book Productions, Berkshire House, Queen Street, Maidenhead, Berks., SL6 1NF in conjunc'ion with Seymour Press Ltd.

1st edition, 1st impression 2.76.0

Made and printed by C. Nicholls & Company Ltd

ISBN 0 85047 907 X

Contents

Introduction

The *Key Facts Physics Passbook* provides a common core of the subject-matter which the UK examining boards require students to know for the purposes of the GCE O-level and CSE physics examinations. Well before taking the examination, students are advised to obtain a copy of the syllabus issued by their chosen examining board to ensure that all the topics listed have been covered.

The chapters of this book are by no means mutually exclusive. There is a considerable overlap of ideas in the various areas of physics, just as there is in any science subject, so it is worth bearing in mind that the examination questions to be answered are likely to include matter from two or even three of these chapters. (For example, this is particularly common in the case of questions on energy changes.)

At the end of each chapter you will find, firstly, a list of key terms occurring in the chapter, and secondly, a number of worked examples illustrating the type of question to be regularly found on physics papers.

Passing examinations requires not only a thorough knowledge of the subject-matter, but also of how to supply answers in the way in which the examiner wants them. The section on examination hints should help you to develop that celebrated skill, 'examination technique'.

In the Key Facts Revision Section are listed in as concise a form as possible all the essential laws and definitions: these should be learned very thoroughly, for they provide the framework of good, clear answers. If these fundamental principles are not fully understood, marks cannot help but be lost.

Hopefully this book will help you to understand your coursework and later, to revise methodically, so that you finish up with a physics qualification of real worth and usefulness to your future.

Chapter 1
Properties of Matter

Solids, liquids and gases

Many **solids** are made up of a **crystal structure**, either a single crystal as in salt and sugar and copper sulphate or **polycrystalline** as in metals. The single atoms or small molecular groups form themselves into a regular pattern to build up the basic crystal shape (figure 1).

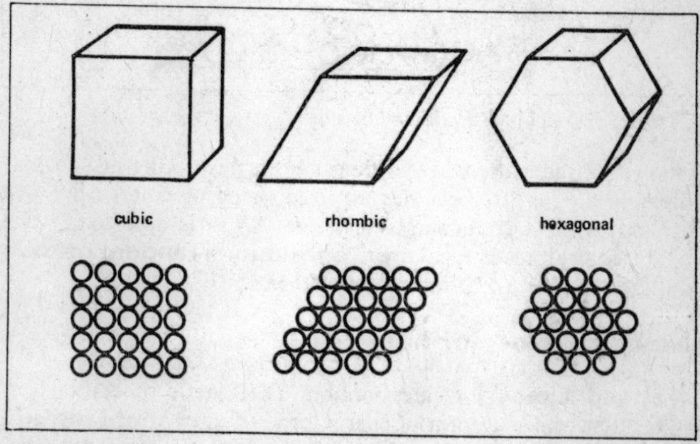

Figure 1. Three crystal shapes

Other solids are made up of long, thin molecules arranged in a random, criss-cross fashion. Plastics, nylon and other man-made materials are included in this category.

Most solids, particularly the crystalline ones, will change into liquids when heated to a certain temperature. When they do so there is only a **slight volume change**, usually an increase. Ice is an exception: when it changes into water its volume decreases.

Liquids have a **fixed volume**, like solids, but **no fixed shape**. They have certain properties called **surface tension** effects that can be explained if the atoms in the liquid are assumed to exert attractive forces upon each other.

Liquids, when heated, change into gases. There is a large volume

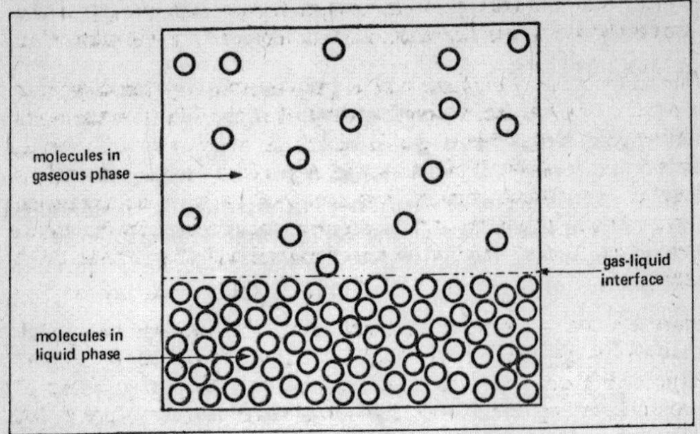

Figure 2. Model of a liquid and its gas

change (of the order of 1 000 times) when this occurs at constant pressure (figure 2). A gas has no fixed shape or volume but will expand to fill a container of any size. A gas is able to do this because the atoms in a gas are in **continuous random motion** not exerting any attraction on each other.

Kinetic theory of heat

The word 'kinetic' indicates motion. The kinetic theory of heat states that heat is a form of energy producing **random motion of the atoms or molecules** of the body that contains the heat. When a substance is heated, the random motion of the atoms increases. In a solid this random motion is a **vibration** about a fixed point in the solid structure. This is why a solid expands when heated: the atoms need a little more space in which to vibrate more rapidly.

When a solid changes to a liquid, **energy is needed to break down the crystal structure**. This energy is the **latent heat of fusion** (see page 69). When a liquid is heated the atoms or molecules move faster, the liquid expands and becomes less viscous. Individual atoms in a liquid will not all have the same speed and energy: some will move faster and some slower than average. A faster-moving atom moving near the surface may have enough energy to escape the attraction of neighbouring atoms and leave the body of the liquid, entering the gaseous state. This explains how evaporation takes place, and may do so even at low temperatures.

If only the high-energy atoms escape, the average energy of the remaining atoms will decrease: a fall in temperature will show this.

The kinetic theory explains why **a gas exerts a pressure**; it is caused by the **random bombardment** of the atoms on the walls of the container. When a gas is heated its pressure or its volume, or both, increase(s). If, for example, a gas in a sealed container is heated, its volume remains constant but its pressure increases. According to kinetic theory, the pressure rise is caused by the atoms moving faster and thus hitting the container walls harder and more often.

Simple models illustrating the kinetic theory of gases are very useful. A two-dimensional model could be a tray of marbles. If the tray is jerked about on a flat table the marbles will move about at random, hitting each other and the sides of the tray (figure 3a). Another model might be a perspex tube containing small metal balls (figure 3b). The balls are energised by a vibrating piston at the bottom. The speed of vibration of the piston and hence the **energy of the balls** can be varied. This is the equivalent of **changing the temperature** of the gas. In these two models, the marbles and the metal balls represent atoms (or molecules) of a gas.

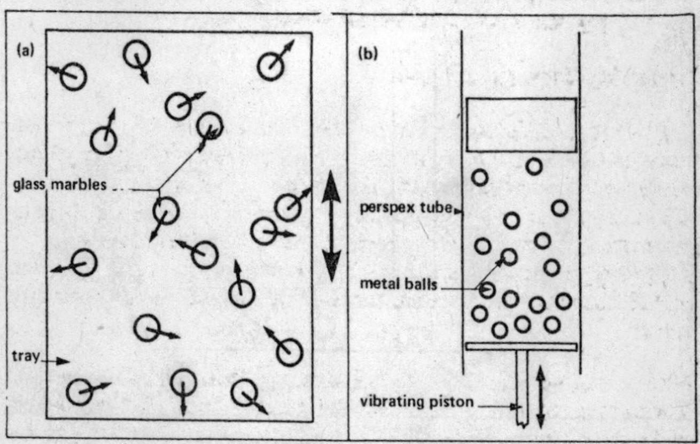

Figure 3. Kinetic models of a gas

These models have their limitations, of course. Energy has to be put into the models continuously, and the balls lose energy each time they hit each other and the walls. No such energy is lost in a gas. There being **no loss in energy** when atoms in a gas collide, it may be concluded that atoms are perfectly elastic. The models' other limitation is the size and number of marbles and balls: there are many, many more molecules, all very much smaller, in a gas. However, the models do show the effect of individual balls striking the walls. To see the effect with real gas a microscope must be used.

Brownian motion

In 1827 a botanist, Robert Brown, was looking at pollen grains through a microscope. He saw them moving in a continuous, haphazard motion. He could not explain this motion at the time, but thought it was caused by the 'life-giving force' in the pollen.

Similar motion is produced in smoke particles. It can be explained by the **random bombardment of the smoke particles by the air molecules**. The smoke particles are small enough to be affected by the air molecules and large enough to be seen under a normal school microscope (figure 4).

The two models described above can be used to show the effect either by placing a beer-bottle top on the marble tray or a polystyrene ball in the perspex cylinder.

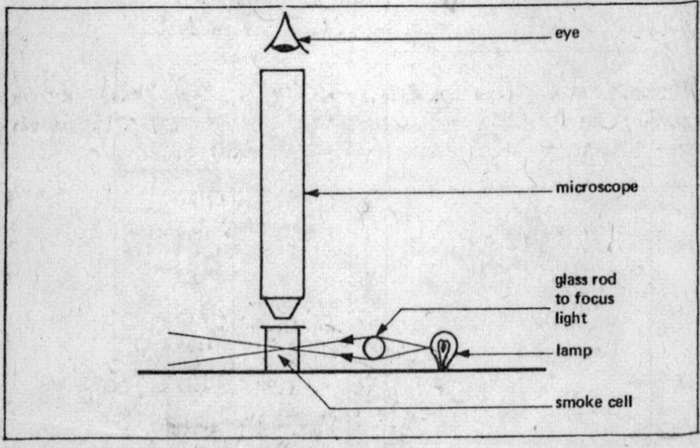

Figure 4. Smoke cell for observing Brownian motion

Oil film measurement

The theory of Brownian motion gives direct evidence for the kinetic theory of gases. It can be used in a quantitative way to explain the gas laws. By measuring the density and pressure of the air, the average **speed of air molecules** can be found—about **500 m/s**.

To gain some idea of the size of molecules, the **thickness of an oil film** can be evaluated.

An oil molecule is long and thin. One end is attracted to water and the other end is repelled by water but attracted by other oil molecules. Thus when a small oil drop is put on to the surface of water it will spread out till it is just one molecule thick.

If a small drop is obtained on a V-shaped piece of thin wire, by using a hand lens and a half-millimetre scale the diameter of the drop can be estimated (usually about 0·5 mm).

If the drop is then placed on a clean water surface that has a light dusting of powder on it, the oil will spread out, pushing the powder away, into a circle (figure 5). The diameter of the circle should be measured (usually about 20 cm).

In the calculation the volume of the drop is equal to the volume of the film. The final answer is best expressed in metres, so the diameter of the drop and the diameter of the oil circle must also be in metres. So let *d* be the thickness of the oil film.

$$\frac{4}{3} \times \pi \times \left(\frac{0{\cdot}5}{2} \times 10^{-3}\right)^3 = \pi \times \left(\frac{20}{2} \times 10^{-2}\right)^2 \times d$$

This indicates a value for *d* of about $\mathbf{2 \times 10^{-9}}$ **m**. This is a very rough method, but a good one for giving the first indications of the size of molecules. (Single atoms will be smaller still.)

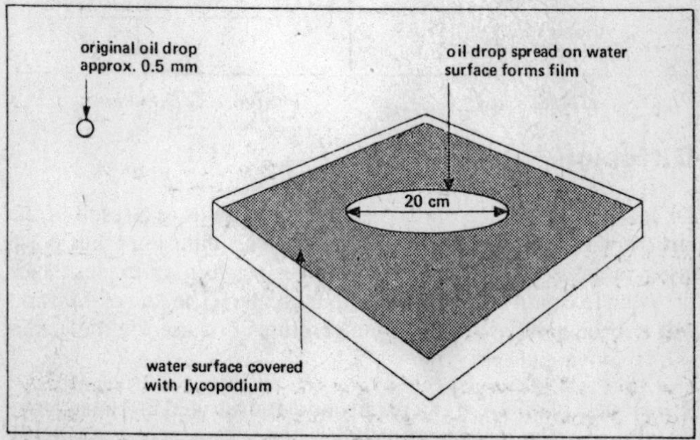

Figure 5. Oil film experiment

Elasticity and Hooke's law

Elasticity is the ability of a substance to recover its original shape after distortion. **Hooke** discovered the simple relationship between the extent of deformation and the force producing it: **provided the elastic limit is not exceeded, the deformation of a material is proportional to the force applied to it.**

For a spring or wire, if a graph of load is plotted against the extension produced, **a straight line passing through the origin will result** up to the elastic limit (figure 6). If the load is removed at any time up to this point the spring will return to its original length. Loading past the elastic limit increases the extension considerably and permanently deforms the spring.

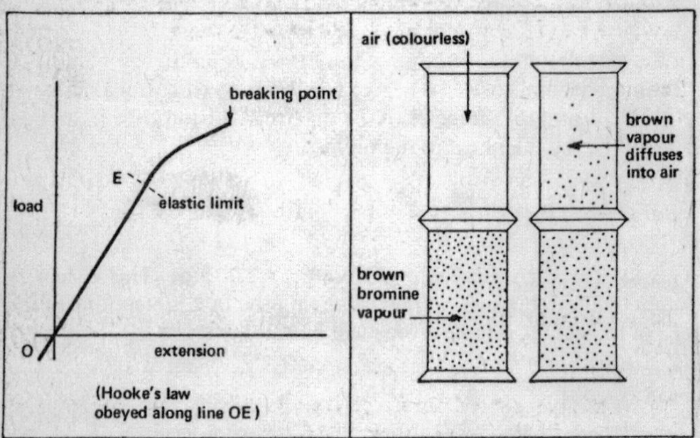

Figure 6. Hooke's law

Figure 7. Diffusion of gases

Diffusion

Diffusion is the mixing of two gases or two liquids as a result of the **random motion of the molecules.** This diffusion takes place faster in gases where there is more space between molecules (since the mean free path between collisions is greater). The rate of diffusion will also be greater at higher temperatures because the molecules will be moving faster.

The process of diffusion can readily be demonstrated in a laboratory by observing the movement of bromine gas in a gas jar (figure 7). A gas jar full of bromine vapour is placed in the fume cupboard and an empty gas jar placed over it as shown. The brown vapour is seen to diffuse slowly into the air in the second gas jar.

Pressure

The word 'pressure' has a definite scientific meaning and should not be used thoughtlessly when 'force' would be more appropriate.

$$\text{Pressure} = \frac{\text{force}}{\text{area}} \qquad \text{S.I. units: newtons/metre}^2$$

A commonplace example of the use of pressure is pressing in a drawing-pin. If you press on the head of a drawing-pin with a force of 4 newtons the pin point will push on to the drawing-board with the same force. However, the pressure produced by the pin point is considerable because of its small area. Let us say that the flat head has an area of 2 cm^2 and the pin point is 0·5 mm^2.

$$\text{Pressure on thumb} = \frac{4\text{N}}{2\ \text{cm}^2}$$

$$= 2 \times 10^4\ \text{N/m}^2$$

$$\text{Pressure on board} = \frac{4\text{N}}{0{\cdot}5\ \text{mm}^2}$$

$$= 8 \times 10^6\ \text{N/m}^2$$

The pressure at the pin point is 400 times greater than at the pin head.

Pressure in a liquid increases with depth and with the density of the liquid. It will also depend on the pull of gravity.

Pressure = depth × density × acceleration due to gravity.
Because the molecules in liquids and gases move in all directions they will produce a force in a direction at right angles to any solid surface in them, upwards and sideways as well as downwards.

Mercury barometer

This is an instrument for measuring **atmospheric pressure**. It was first suggested by **Torricelli**. A strong glass tube about 90 cm long, closed at one end, is filled with mercury. The open end is then blocked and lowered into a bowl of mercury. Then the open end is unblocked. The mercury falls away from the closed end of the tube till the height of the top is about 76 cm from the mercury level in the bowl. The mercury is kept at this height by the atmospheric pressure (figure 8a).

The space above the mercury is empty and thus exerts no pressure on the top surface of the mercury. (More correctly, there is mercury vapour but its pressure is negligible and is usually ignored.) The pressure exerted by the mercury (with a height of 76 cm) can be calculated using the formula stated below. This will indicate the pressure of the air.

$$\text{atmospheric pressure} = \text{height of column of mercury} \times \text{density of mercury} \times g$$

$$\text{height of mercury column} = 0{\cdot}76 \text{ metres}$$

$$\text{density of mercury} = 1{\cdot}36 \times 10^4 \text{ kg/m}^3$$

$$g \text{ (pull of gravity)} = 9{\cdot}8 \text{ m/s}^2$$

$$\therefore \quad \text{atmospheric pressure} = 0{\cdot}76 \times 1{\cdot}36 \times 10^4 \times 9{\cdot}8$$

$$= 1{\cdot}013 \times 10^5 \text{ N/m}^2.$$

As the atmospheric pressure changes, the height of mercury will increase or decrease.

Note The height measured is the **vertical distance**; if the barometer is tilted the height will remain the same (figure 8b).

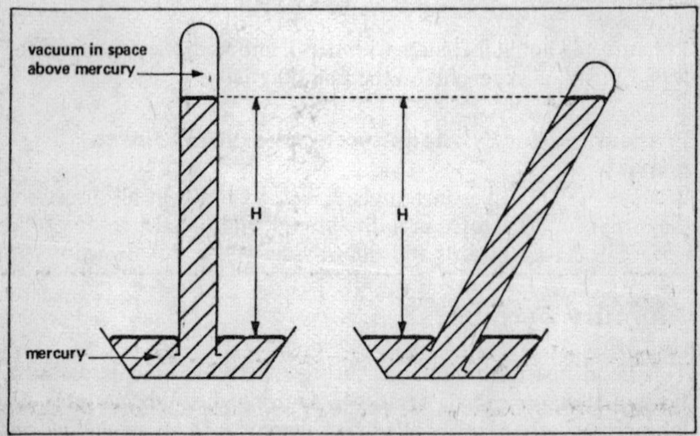

Figure 8. Mercury barometer: (a) upright; (b) tilted

U-tube manometer

If water is poured into a U-tube it will always settle with the levels in both arms the same. Both sides are open to the air so the pressure on both surfaces is the same. If one arm is connected to the gas supply the height of the water level will alter. The level in the side open to the atmosphere will go up and the level in the other, connected to the gas supply, will go down. The excess pressure of the gas over atmospheric pressure will balance the excess height of the water column in the open arm (figure 9). The excess pressure of the gas may be stated in centimetres of water or the pressure calculated in N/m^2 in the same way as for the pressure of the column of mercury in a barometer (see above).

Water is a suitable liquid to use where small differences in pressure are to be measured. To measure greater pressure differences, mercury is used; because of its much greater density it can be used in a correspondingly smaller piece of apparatus than a water U-tube.

Note U-tubes measure only the **difference in pressure between the two surfaces.** These can be less or greater than atmospheric pressure.

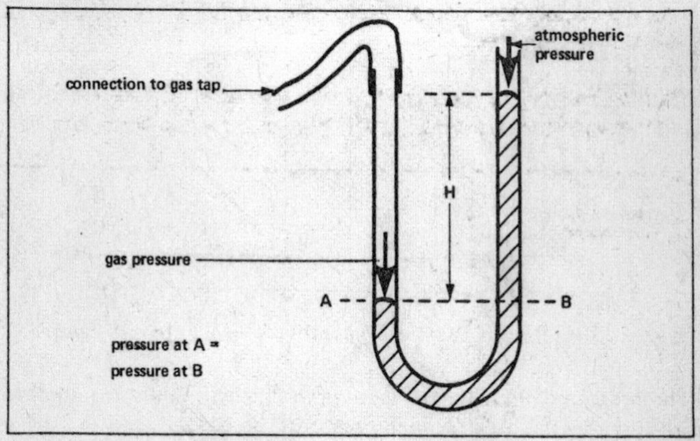

Figure 9. U-tube manometer

Practical application of liquid pressure

Hydraulic systems for transmitting force and energy have a wide application in engineering. In such hydraulic systems, pressure is applied to a fluid by exerting force on a piston. The pressure is then transmitted by the fluid, and may be used to apply a force to a second piston in the system. An example of this is the hydraulic braking system used in most cars.

Boyle's law

Using a bicycle pump is a simple demonstration of the fact that when the volume of a gas is decreased its pressure increases. This effect was first accurately measured by **Robert Boyle**, who found a simple **inverse relation between pressure and volume.**

Certain precautions have to be taken to **keep the temperature of the gas constant**. The experiment is usually carried out at room temperature, which is assumed to be constant. The apparatus used is shown in figure 10a. The volume of gas is proportional to the height (H) of the gas column (provided the cross-section of the tube is uniform).

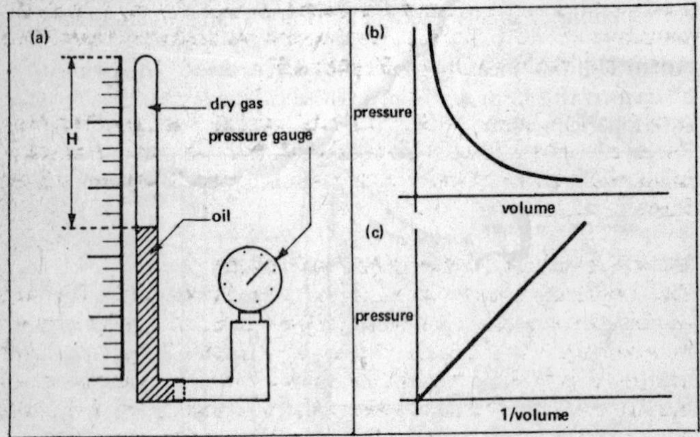

Figure 10. (a) Boyle's law apparatus; (b) p/V graph; (c) $p \Big/ \frac{1}{V}$ *graph*

The pressure/volume graph is a curve that will never cross either axis (figure 10b). At very high pressures the gas will still occupy a small volume and if the gas is expanded to a very large volume it will still exert a small pressure.

The pressure$\Big/\frac{1}{\text{volume}}$ graph is a straight line through the origin (figure 10c). Also for any pair of values of pressure × volume the same number is obtained.

This verifies the law that Boyle first suggested:

The volume of a fixed mass of gas is inversely proportional to the pressure, provided the temperature remains constant.

Flotation and Archimedes' principle

When a solid body is situated in a liquid or a gas, an upthrust is exerted on the body which makes the body appear lighter. In fact sometimes the body floats, and thus appears to weigh nothing at all, e.g. a piece of wood floating on water, or a balloon floating in the air.

The amount of upthrust depends upon two factors, **the volume of the body and the density of the fluid**. Archimedes was the first to state that **when a body is wholly or partially immersed in a fluid it experiences an upthrust equal to the weight of the fluid displaced**.

A body will float in a fluid if its total, overall density is less than the density of the fluid. A steel ship will float on water because it is hollow. A block of lead will float on mercury. A hot-air balloon floats in cold air.

Verification of Archimedes' principle

The apparatus shown in figure 11 can be used to illustrate Archimedes' principle. A solid (e.g. a lump of lead) is weighed in air by attaching it to the hook of a spring balance. The solid is then immersed in water contained in an overflow can and the water displaced is collected in a beaker. The new reading of the spring balance is noted. It is found that the difference in weights of the solid in air and in water equals the weight of water displaced. The density of water is 1 g/cc hence the volume of water displaced (in cc) is numerically equal to the weight of that water (in grams).

Density

Which is heavier—a kilogram of lead or a kilogram of feathers? They are both the same, of course; but what the non-scientist means when he says that lead is heavier than feathers is that if **equal volumes of each were considered**, the lead would indeed weigh more than the feathers. A scientist would say that **lead has a greater density than feathers**.

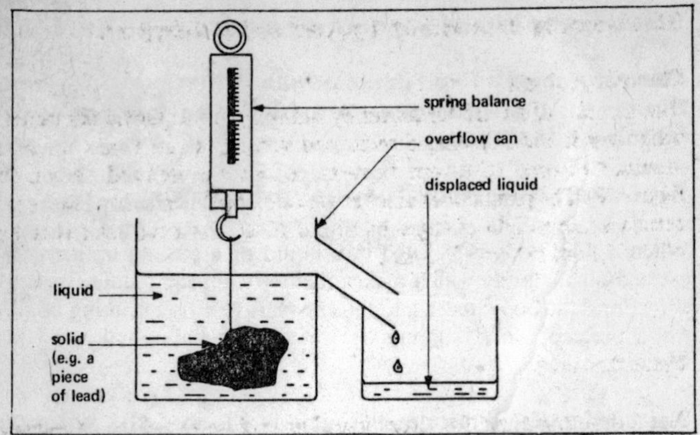

Figure 11. Apparatus to illustrate Archimedes' principle

Density means the mass per unit volume of the substance.

$$\text{Density} = \frac{\text{mass}}{\text{volume}}$$

Unit of mass = kilogram, unit of volume = metres^3, thus density unit = **kilograms/metres^3**.

In some school laboratories there is still a tendency to use grams instead of kilograms and centimetres instead of metres; in this case the unit of density will be **g/cm^3**.

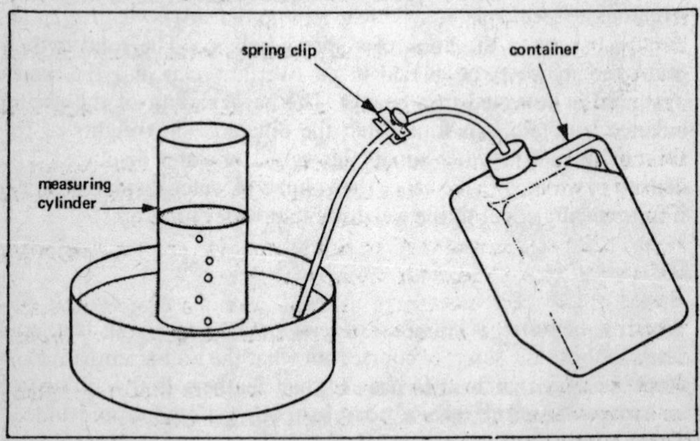

Figure 12. Apparatus to determine the density of air

The density of water is 1 g/cm^3 or 1 000 kg/m^3.

Density of air

The density of air can be found by pumping up a plastic container, weighing it and allowing a measured volume of air to escape (the volume allowed to escape is measured by the method shown in figure 12). The plastic container is then weighed. For every 1 000 cm^3 released the weight changes by about 1·2 g. Thus air has a density of 0·001 2 g/cm^3 or 1·2 kg/m^3.

Key terms

Surface tension An effect found at a liquid surface. It causes the liquid surface to behave as if it were a stretched skin. It may be explained by the fact that liquid molecules attract each other.
Kinetic theory of matter Matter is made up of particles that are in continuous random motion.
Brownian motion The term given to the random motion of particles (e.g. pollen grains) that can be observed with the aid of a microscope. The particles are large enough to be seen with a microscope but small enough to be set in motion owing to their being bombarded by the molecules of the fluid surrounding them.
Elasticity The ability of a substance to recover its original shape after distortion.
Hooke's law Within the elastic limit the extension is proportional to the force producing it.
Diffusion The mixing together of two gases owing to the random motion of the molecules.
Pressure The force acting on unit area, stated in units of newtons/metre^2 (N/m^2).
Barometer An instrument for measuring atmospheric pressure.
U-tube manometer An instrument for measuring pressure differences. Usually one side is open to the atmosphere.
Hydraulic systems Used in engineering for transmitting force and energy through liquids contained in pipes.
Boyle's law The volume of a fixed mass of gas is inversely proportional to the pressure, provided the temperature remains constant.
Archimedes' principle A solid immersed in a fluid experiences an upthrust equal to the weight of the fluid it displaces.
Density The mass per unit volume, stated in units of kilograms per cubic metre (kg/m^3).

Examples

Archimedes' principle

A uniform wooden cube of side 1 m and mass 600 kg floats in a liquid of density 800 km/m^3. What is the weight of the wood? What additional force would be required just to submerge the wood? (Take $g = 10$ m/s^2.)

This is a question about forces. If an object is floating in a liquid, then the downward force (i.e. the weight of the object) must be balanced by an equal upward force (upthrust). If the downward force is greater, the object will sink, and if the upward force is greater, the object will rise higher in the liquid.

Remember that Archimedes' principle states that the upward force is equal to the weight of fluid displaced.

The weight of the wooden cube is the product of its mass and the acceleration caused by gravity, i.e.
weight $= mg = 600 \times 10 = 6\,000$ N.

If the wood is just submerged, all of it will be under the surface of the liquid. The volume of the wood is $1 \times 1 \times 1 = 1$ m^3. This means that 1 m^3 of liquid will be displaced by the wood. Since the density of the liquid is 800 kg/m^3, the mass of liquid displaced is 800 kg. The weight of this displaced liquid is mg which in this case is $800 \times 10 = 8\,000$ N, and this is equal to the upthrust.

So if the block of wood is just submerged, it has a downward force of 8 000 N acting on it equal to the upward force of 8 000 N. But the downward force caused by its weight is only 6 000 N, therefore an additional downward force of $8\,000 - 6\,000 = 2\,000$ N must be applied to hold it in position.

Hydraulic systems

A small piston of cross-section area 5 cm^2 carries a load of 8 kg. It is connected by a hydraulic system to a large piston of area 500 cm^2. What force is exerted by the large piston?

The pressure in the liquid $= 8 \times 10$ N/5 cm^2 $= 16$ N/cm^2.
Force on large piston = pressure $\times$ area $= 16 \times 500$ N $= 8\,000$ N.
Thus load of piston = 800 kg.

Boyle's law

A cylinder contains a fixed mass of gas enclosed by a piston. The volume of gas is initially 0·25 m^3 and its pressure is 10 N/m^2. The volume is then reduced to 0·1 m^3 at constant temperature. Calculate the new pressure of the gas and explain whether energy has to be supplied or removed during the process in order to keep the temperature constant.

The calculation is a simple application of Boyle's law. We know $p_1 = 10\ N/m^2$, $V_1 = 0{\cdot}25\ m^3$ and $V_2 = 0{\cdot}1\ m^3$, so:

$$10 \times 0{\cdot}25 = p_2 \times 0{\cdot}1$$

$$\therefore \quad p_2 = 25\ N/m^2$$

The second part of the question is rather more involved. We can argue that the piston has to be pushed in in order to compress the gas. In other words, a force must be applied to the piston. This force moves through a certain distance as compression occurs, so work is done. Work requires energy, so energy has been supplied to the system, which will cause an increase in temperature. Hence, in order to keep the temperature constant, energy will have to be removed.

Mercury barometer

A barometer reads 0·766 m of mercury at the bottom of a building. Explain what happens to the reading of the barometer when it is taken to the top of the building, and calculate the height of the barometer there if the building is 60 m high.
The densities of air and mercury are 1·29' and 13 600 kg/m^3 respectively.

The barometer will register a lower reading when taken to the top of the building. The barometer measures atmospheric pressure, and this is caused by the weight of the column of air acting downwards on the earth. This column of air is shorter at the top of the building than at the bottom. The pressure is a measure of the weight of the column of air acting on unit area, and so is less at the top of the building than at the bottom.

Pressure at top of building = pressure at ground − pressure resulting from 60 m of air.

Pressure resulting from 60 m of air = $60 \times g \times 1{\cdot}29\ N/m^2$
If h is the depression of the mercury column, then pressure resulting from h m of mercury = $h \times g \times 13\,600\ N/m^2$

$\therefore \quad 60 \times g \times 1{\cdot}29 = h \times g \times 13\,600$

Note that g cancels out, so

$h = 60 \times 1{\cdot}29/13\,600 = 0{\cdot}005\,69$ m, or $0{\cdot}006$ m

$\therefore$ barometer reading $= 0{\cdot}766 - 0{\cdot}006 = 0{\cdot}760$ m.

Chapter 2
Motion

Uniform velocity

'Velocity' has a slightly different meaning from 'speed' in that velocity is a speed in a definite direction, either stated or implied. Both can be defined as **distance travelled in unit time**. The word **displacement** is used to indicate distance in a specified direction.

$$\text{Velocity} = \frac{\text{displacement}}{\text{time}}$$ Units will be metres/seconds (m/s).

Acceleration

A train travelling at 50 km/hr increases its velocity to 70 km/hr in 5 minutes.

Its acceleration is $\frac{(70-50)\text{ km/hr}}{5\text{ minutes}} = 4\text{ km/hr/minute}$

Its speed has increased by 4 km/hr per minute.

Note that in the units of acceleration there are **two units of time**. In the above example they are hours and minutes. Usually the same unit is used twice.

$$\text{Acceleration} = \frac{\text{change in velocity}}{\text{time taken}}$$

Graphs

Distance/time graphs

A graph in which a straight line is drawn through the origin indicates uniform velocity (figure 13a).

A curved line means a changing velocity: a curve upwards indicates an increase in velocity, an acceleration (figure 13b); a curve downwards indicates a decreasing velocity, a negative acceleration (figure 13c).

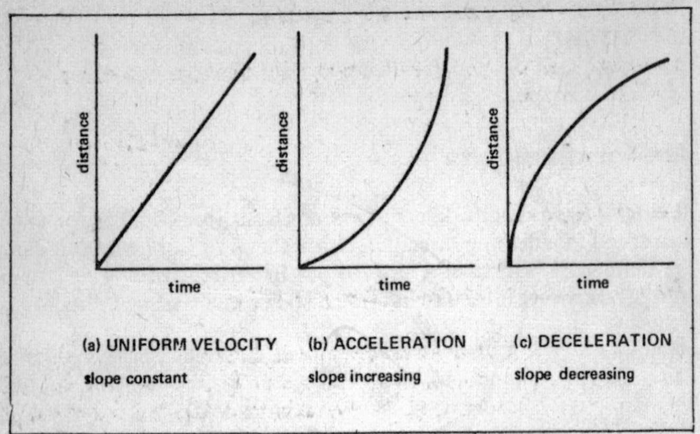

Figure 13. Distance/time graphs

Velocity/time graphs

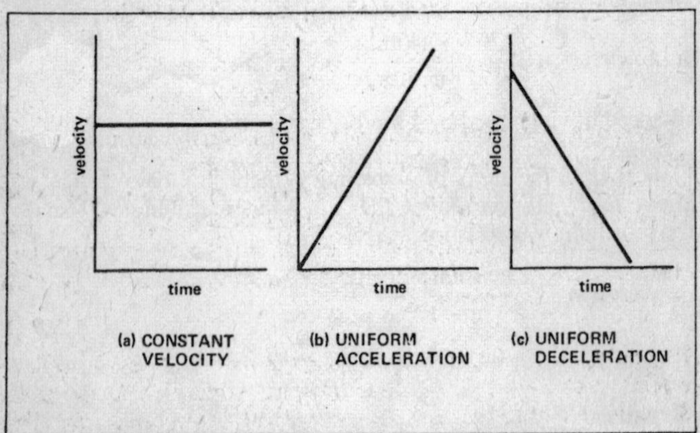

Figure 14. Velocity/time graphs

A horizontal straight line represents a constant velocity (figure 14a). A straight line through the origin represents uniform acceleration starting from rest (figure 14b). A straight line sloping downwards indicates a negative acceleration (figure 14c).

The distance travelled by the object is represented by the area under the velocity/time graph.

The four equations of motion

u = origin velocity; v = final velocity; s = distance; t = time; a = acceleration.

$$v = u + at \qquad s = \frac{(v+u)t}{2}$$

$$s = ut + \tfrac{1}{2}at^2 \qquad v^2 - u^2 = 2as$$

Acceleration caused by gravity

All objects will accelerate to the ground if dropped. If the effect of air resistance is ignored, then bodies of differing weights will all have the same acceleration. **A stone and a feather will drop together in a vacuum.** Acceleration caused by gravity is denoted by g.

Experimental method for finding *g*

For a body dropping from rest the equation of motion is:

$$s = \tfrac{1}{2}gt^2 \quad \text{or} \quad g = \frac{2s}{t^2}$$

Most school laboratories have millisecond timers that can be started and stopped electrically. One can be used to find the time a steel ball takes to drop through a measured distance. The ball itself breaks an electrical contact at the start of the fall to start the timer, and hits a switch at the end of its fall to stop it. A number of readings should be taken for various heights and the value of g calculated from the formula.

For most calculations the value of g can be taken as 10 m/s². A more accurate value is 9·81 m/s². It varies from place to place on the earth's surface because the earth is not exactly spherical. The value of g is higher at the poles and lower at the equator.

Force

A force is a push or a pull. One force acting on a body will cause it to accelerate. If there are two or more forces acting on a body and the forces balance, the body may be made to change its shape, e.g. stretch.

A series of experiments can be performed with ticker-tape attached to trolleys (see below). These show that **the greater the force acting on the trolley the greater is its acceleration. If the mass of the trolley is increased the acceleration will be less.**

This law provides a method of defining the unit of force:

force = mass × acceleration **newtons = kg × m/s^2**

Experiment with trolleys and ticker-tape to show relationship between mass, acceleration and force

A vibrator is connected to a low-voltage alternating supply that makes it vibrate 50 times a second. A paper tape attached to the trolley passes under the vibrator which imprints dots on the tape (figure 15a). The time interval between one dot and the next on the tape will therefore be 0·02 seconds.

To obtain a constant force (F), an elastic band attached to the trolley is pulled to a **fixed length**. Forces of value $2F$ and $3F$ are obtained by using two and then three elastic bands. For $2M$ and $3M$, first two and then three trolleys are put on top of each other. The tape is cut into five-gap lengths, each length representing the distance travelled in 0·1 second (figure 15b). The size of the steps between neighbouring five-gap lengths represents the acceleration produced. The trolley runway is sloped slightly to compensate for friction. When the force is doubled, the difference in lengths of successive five-gap lengths is doubled. If the mass of the trolley is doubled, the difference in length between successive five-gap lengths is halved.

Weight

Weight is the pull of gravity on a body. A weight of 1 kg is approximately 10 newtons, but it will vary from place to place, being greater at the poles and less on the equator.

The weight of a body on the moon will be much less, about 1·8 newtons to each kg. Hence, a man with mass 200 kg will weigh 2000 N on earth but only 360 N on the moon.

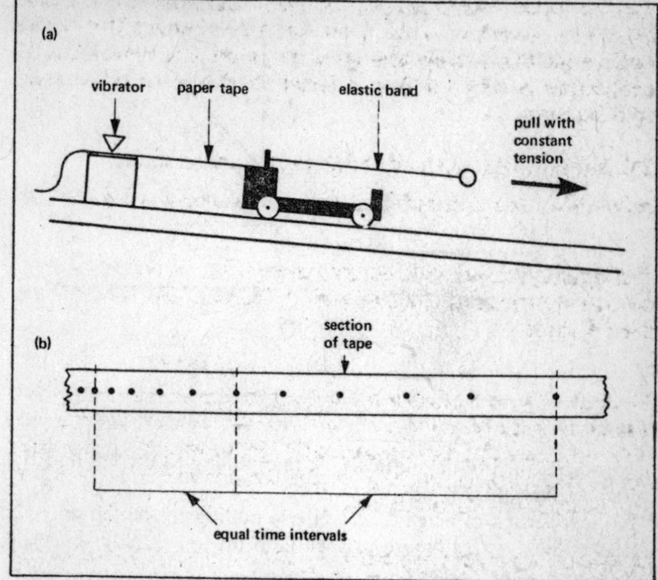

Figure 15. Accelerating trolley experiment

This gives us another way of measuring **the strength of a gravitational field**, not in terms of the acceleration it will bring to bear on a freely-falling body but in terms of **the force acting on 1 kg**. In fact both these definitions lead to the same numerical value:
a field of 10 N/kg will produce an acceleration of 10 m/s^2 on any freely-falling body.

Vectors

Force, velocity, acceleration and momentum are all **vector** quantities, that is they all have **magnitude and direction.**

Normal scalar arithmetic cannot be used for adding vectors that are not both in the same direction. In vector arithmetic $\vec{1}+\vec{1}$ can equal $\vec{2}$ or $\vec{0}$ or anything between, depending on the directions. (The arrows remind us that we are dealing with vectors.)

Thus two forces each of magnitude 1 N both pulling to the left will add to give 2 N to the left.

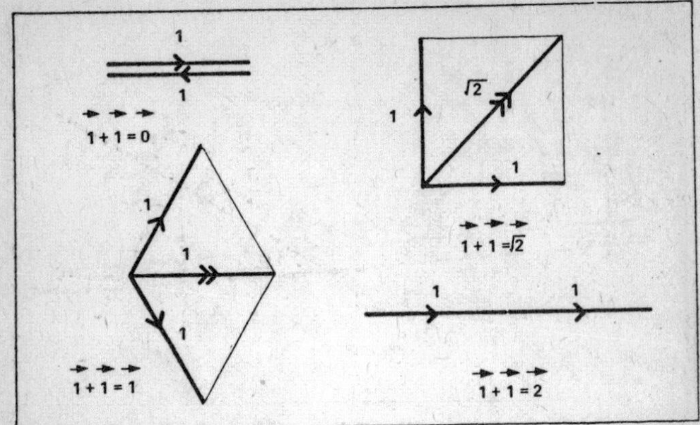

Figure 16. Vector addition

Two forces each of magnitude 1 N, one pulling to the left and the other to the right, will add up to zero.
Two forces each of magnitude 1 N one pulling to the left and the other pulling vertically upwards will add up to $\sqrt{2}$ N at an angle of 45° to the horizontal.

In more general cases each vector is represented by a line, the length of the line representing the magnitude of the vector, and its direction on the paper indicating the direction of the vector. The two vectors to be added are drawn from a single point, a **parallelogram** is constructed and the **resulting diagonal** denotes the vector sum in both magnitude and direction (figure 16).

Consider the problem of an aircraft flying due north through the air at 500 km/hr and the air moving due east at 200 km/hr. The magnitude and direction of the plane relative to the ground can be found by drawing a vector diagram (figure 17a).

Relative to the ground the plane is flying at 540 km/hr on a course N 22°E.

Resolving a vector into two components at right angles

The reverse operation can be carried out. A single force of 10 N pulling down on 1 kg mass that is resting on a slope at 30° to the horizontal can be considered as 5 N down the slope and 8·7 N perpendicular to the slope (figure 17b).

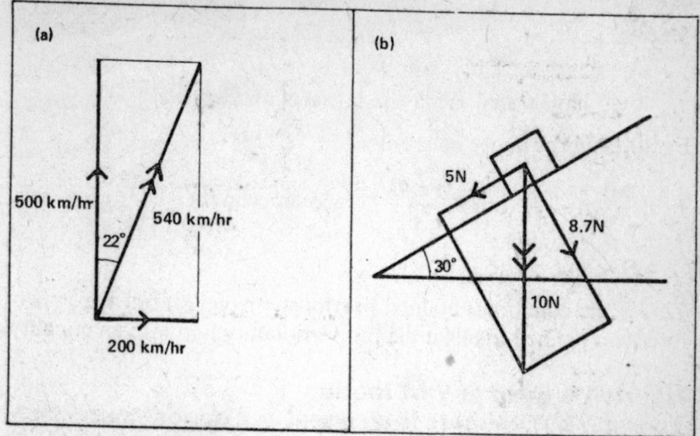

Figure 17. Parallelogram for (a) vector addition; (b) resolution

Newton's laws of motion

From the formula 'force = mass × acceleration' it is clear that if there is no force acting on a mass there can be no acceleration. This is **Newton's first law**, which stated formally is: **every body continues in its state of rest or uniform motion in a straight line unless there is an unbalanced external force acting on it**.

A book lying on a table has two forces acting on it: the pull of gravity downwards and the reaction of the table upwards. These two forces are equal and opposite, therefore the book remains at rest. An express train moving at 150 km/hr also has no unbalanced forces acting on it: gravity downwards is balanced by the upward push of the rails, and the frictional resistance of the air and the wheel bearings is balanced by the push of the driving wheels on the rails.

Momentum: Newton's second law

Newton expressed his second law in terms of momentum. To calculate the **momentum** of a body **multiply mass × velocity**.

The rate of change of momentum of a body is proportional to the unbalanced force acting on it and takes place in the direction of the force.

$$F \propto \frac{mv - mu}{t}$$

He could have stated the law in terms of acceleration:

Factorising the above equation:

$$F \propto \frac{m(v-u)}{t} \quad \text{but} \quad \frac{(v-u)}{t} \quad \text{is acceleration } (a)$$

$$\therefore \quad F \propto ma$$

This is the equation obtained previously (page 28) but the unit of force was defined to make the proportionality sign into an equality.

Newton's third law of motion

To every action there is an equal and opposite reaction.

Consider again our book on the table: the book is pushing down on the table with an equal and opposite force to the table pushing up on the book.

In the express train example, the rails are pushing the train forwards and the train is pushing the rails backwards.

Consider a bullet being fired from a gun: the gun pushes the bullet forwards, the bullet pushes the gun backwards.

This is the principle of the rocket engine and explains why it can function out in space with nothing to push against. The rocket pushes the hot gases out at the back and the gases push the rocket forward with the same force.

Law of conservation of momentum

This is one of the four conservation laws of physics, the other three being the **conservations of angular momentum**, of **electric charge** and of **mass-energy**. The latter was considered two separate laws until the spitting of the atom early in this century.

The conservation of momentum follows as a consequence of Newton's second and third laws.

The equation obtained from the second law can be rearranged as follows:

$$Ft = mv - mu$$

Ft is **force × time** and is called the **impulse**. Therefore **the impulse is equal to the change in momentum.**

If two bodies collide or push each other apart they will push against each other with equal and opposite forces for the same time and impart to each other equal and opposite impulses. They will therefore produce in each other equal and opposite changes in momentum. The sum of the momentum of the two bodies will remain the same. If a third of a fourth or any greater number of bodies are involved the same will be true.

In any closed system the total momentum remains constant

Remember that momentum is a vector quantity, because it has both magnitude and direction. If a body is moving from left to right its velocity and momentum are considered positive and if it is moving from right to left its velocity and momentum will be negative.

Consider the recoiling rifle as an example. The rifle has a mass of 2 kg, the bullet a mass of 50 g (0·05 kg). The bullet leaves the rifle with a velocity of 400 m/s. The velocity of recoil (let it be v m/s) of the rifle can be calculated.

Initially, when neither the rifle nor the bullet is moving, the total momentum will be zero. The total momentum after firing will also be zero.

$$2 \times v + 0{\cdot}05 \times 400 = 0$$

$$v = -\frac{0{\cdot}05 \times 400}{2}$$

$$= -10 \text{ m/s.}$$

The velocity of recoil is 10 m/s. The negative sign shows that it is in the opposite direction to the bullet.

For another example of the use of this law, consider the collision of two trolleys.

For a 'closed system' it is assumed that there is no friction with the bench surface. Two types of collision are usually considered, one in which the two trolleys stick together after collision (called **an inelastic collision**), the other when no energy is lost (called **a perfectly elastic collision**). The second case entails knowledge of kinetic energy (see opposite). Let us consider the simpler inelastic collision.

Let the first trolley have a mass of 4 kg, moving to the right with a velocity of 5 m/s. Let the second trolley have a mass of 6 kg moving to the left with a velocity of 2 m/s (−2 m/s must be used in the calculation). Let the common velocity of the trolleys after collision be v m/s.

$$\text{Total momentum before collision} = (4 \times 5) - (6 \times 2) \text{ kg m/s}$$
$$= 8 \text{ kg m/s}$$
$$\text{Total momentum after collision} = (4 \times v) + (6 \times v) \text{ kg m/s}$$
$$= 10v \text{ kg m/s}$$
$$\text{Equating the two:} \quad 10v = 8$$
$$v = 0{\cdot}8 \text{ m/s.}$$

After collision the two trolleys move to the right with a velocity of 0·8 m/s.

Work and energy

'Work' has a precise meaning in science. When a force produces movement, work is done and energy is transferred from one body to another and possibly transformed from one type to another.

Work = force × distance; the distance must be measured in the direction of the force.

Units will be **newtons × metres**. These have a special name, the **joule**, after the scientist who did so much experimental work to establish that heat is a form of energy.

Potential energy

When a mass of 5 kg is lifted from the ground to a height of 2 m work is done on it, energy has been used up by the person or machine doing the lifting and thus energy is stored in the mass.

Gravitational pull on 5 kg is 50 N thus the work done is 50 × 2 J, that is 100 J. 100 J of energy are stored in the mass as potential energy.

Potential energy can also be stored in a spring. When a spring is stretched or compressed, work is done, and an equal amount of work can be done by the spring when it is released.

Similarly, a gas can be pumped into a container to a high pressure. Work is done in pushing the gas in, and an equal amount can be obtained if the gas is allowed to push its way out.

Kinetic energy

For a body to start moving it must first be pushed, so work is done. To bring itself to rest again that body will have to do an equal amount of work. So when it is moving it is said to have energy.

The amount of kinetic energy a body has will depend upon its mass and its velocity. It can be proved that:

Kinetic energy = $\frac{1}{2}$ mass × velocity² $= \frac{1}{2}mv^2$

Thus a body of mass 5 kg moving with a velocity of 4 m/s will have kinetic energy of:

$\frac{1}{2} \times 5 \times 4^2$ J $= 40$ J.

Changes from potential to kinetic energy and *vice versa*

Among many examples of changes of potential energy to kinetic energy and from kinetic energy to potential energy, the most obvious and commonplace is the freely-falling body. (The term 'freely-falling' denotes that air resistance can be ignored.)

Another example is the **pendulum**, representing an almost reversible change from potential at the top of its swing to kinetic energy at the bottom of its swing and back to potential again (figure 18a).

Experiment to measure transfer of potential energy to kinetic energy

Using the apparatus shown in figure 18b, the mass m is raised to a height h where its potential energy is mgh. It is then allowed to fall freely so that its potential energy is transformed into kinetic energy of the trolley and the mass itself. Just before it reaches the ground, the system has a velocity v which can be measured by means of ticker-tape attached to the trolley. The kinetic energy is $\frac{1}{2} \times (M+m) \times v^2$, where M is the mass of the trolley.

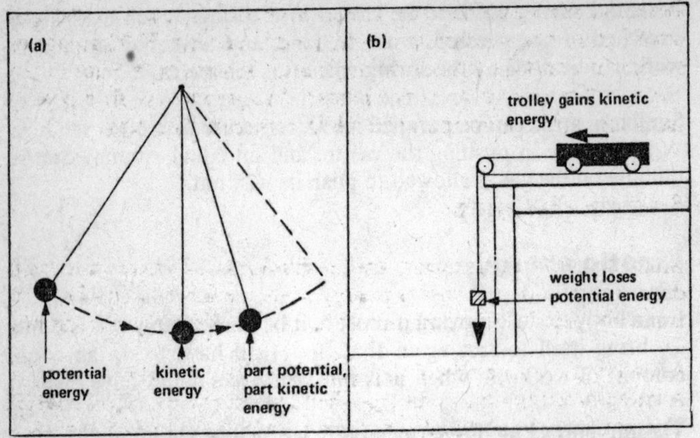

Figure 18. Potential/kinetic energy changes: (*a*) *simple pendulum;* (*b*) *falling mass*

In such an experiment the potential energy will be greater than the kinetic energy owing to the heat energy produced by friction in the trolley wheels, the pulley and the tape.

Other forms of energy

Chemical energy Two or more chemicals react and may at the same time give out heat and light energy, as in a fire, or electric energy, as in a battery.

Heat energy This is the kinetic energy in the random motion of atoms or molecules. To raise the temperature of a body this random motion has to be increased.

Light energy Just as water waves carry energy from place to place, so light waves carry energy from place to place.

Electrical energy It is possible to store only small amounts of electrical energy by charging up capacitors. Electric currents are a most useful means of transferring energy from place to place.

Sound energy Sound is a longitudinal vibration usually transmitted through the air, though it is also transmitted through solids and liquids. Energy is used in setting up the vibrations and is absorbed when the waves are stopped.

Nuclear energy This is the energy source of the sun. A hydrogen bomb explosion is a similar, uncontrolled reaction. Nuclear power stations control the rate of the reaction. Changes take place in the nucleus of the atoms, and one element changes into another. A very small amount of mass is lost which is converted to energy.

Energy changes

Many areas of engineering and technology are concerned with developing ways of changing energy from one form to another and transferring it from one place to another as efficiently as possible.

A few simple examples will illustrate this:
A **candle** is designed to change chemical energy into light energy. The heat energy produced is not required but cannot be prevented.

Similarly in an **electric light bulb** electrical energy is converted into light and the heat energy produced is kept to a minimum
A **fluorescent tube** is more efficient than a filament bulb because it produces more light and less heat.

An **electric motor** changes electrical energy into mechanical energy—either kinetic or potential or both. An **electric dynamo** does the reverse, changing mechanical energy to electrical. In many practical cases the same instrument can be used either as a dynamo or a motor.

An **electric microphone** changes sound vibrations into electrical vibrations. A loudspeaker does the reverse, changing electrical vibrations into sound.

Internal combustion engines, steam engines, jet engines and rockets all attempt as efficiently as possible to change chemical energy into mechanical energy. In all of them a great deal of the energy is wasted as heat.

Solar cells change light to electrical energy.

Thermocouples change heat to electrical energy.

Batteries change chemical to electrical energy.

In **power stations**, heat is produced either by chemical means—oil or coal burning—or by a nuclear reactor. The heat produces steam at high pressure. The pressure turns a turbine that turns a dynamo to produce electricity. In a **hydro-electric power station** the water has potential energy at the top of the dam. As the water rushes downhill, this is changed to kinetic energy, which drives the turbines. The turbines turn the dynamo that will generate the electricity.

Power

One electric motor lifts a load of 5 kg through a vertical distance of 4 m in 10 seconds (s). Another motor lifts 12 kg through 6 m in 30 seconds. Which is the more powerful?

'Power' has a very definite meaning and indeed, it is a word that should never be used in an inexact way. **Power is the rate of doing work**: that is, the amount of work done in 1 second.

For the first motor power $= \frac{\text{work}}{\text{time}} = \frac{50 \text{ N} \times 4 \text{ m}}{10 \text{ s}} = 20 \text{ J/s}$

For the second motor power $= \frac{120 \text{ N} \times 6 \text{ m}}{30 \text{ s}} = 24 \text{ J/s}$

Therefore the second motor is the more powerful.

Note the units for power are J/s, also known as **watts**.

To measure the power of a person working against a brake, as in figure 19, the braking force $(F_1, -F_2)$, and the distance moved

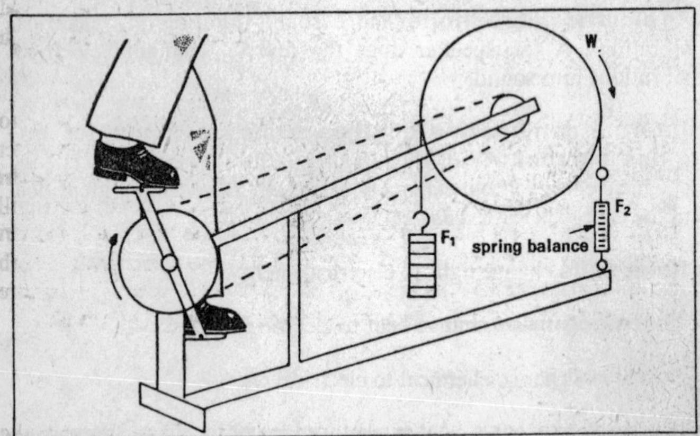

Figure 19. Cyclist working against a brake

against this force in a certain time must be known. The distance is evaluated in terms of the number of revolutions × circumference of the wheel.

Motion in a circle

Newton's first law states that a body will continue to move with uniform speed **in a straight line** unless acted upon by a force. Therefore, for a body to move in an arc of a circle a force must act on it. This force is called **centripetal force**. The direction of this force is **towards the centre of the circle, at right angles to the direction of motion**. Hence this force alters the direction of motion but not its magnitude. This is an example of a force acting but not doing any work because work = force × distance moved in the direction of the force.

The moon moves in a stable orbit around the earth because a centripetal force is produced by the earth's gravitational pull. If the gravitational pull were too great the moon would spiral inwards. If it were too small the moon would spiral outwards. The same conditions apply to the motion of the earth orbiting the sun and with other natural and artificial satellites.

'Weightlessness'

This term is used in two types of situation: (i) if a body is so far away from any other body that it experiences no detectable gravitational pull, i.e. the state of true weightlessness; (ii) the equal and opposite reaction to a man's weight prevents him from falling in normal circumstances.

If, however, a man is enclosed in a large box which is allowed to fall with an acceleration g, both the man and the box will be subject to the same accelerating force. There will therefore be **no reaction to his weight** owing to the floor of the box, and the man will think himself weightless inside it. This is the condition of a man in a spacecraft orbiting the earth. The craft and the man inside it both experience the same accelerating force directed towards the centre of the earth (figure 20). The man is 'weightless' inside the craft.

Friction

In the experiment for measuring the frictional force between the block and the bench (figure 21) the pan is loaded till the block just

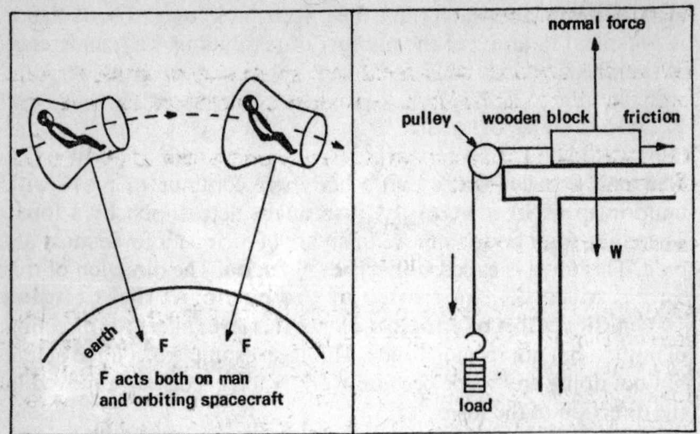

Figure 20.
Orbiting spacecraft

Figure 21.
Experiment for measuring friction

begins to move. The experiment is repeated with blocks of different masses and surface areas. The results can be summarised as follows:
1) The frictional force between two surfaces is proportional to the normal force between the two surfaces. (In this case, the force is the weight of the block.)
2) The frictional force does not depend on the area of the two surfaces in contact.

For many physics experiments every effort is made to reduce friction by using trolleys with ball-bearing axles.

Engineers are often required to reduce friction at one particular point in their machines, but to produce a large amount of friction at another. For example, designers of motor cars do their best to reduce friction in the engine and to have very little rolling friction between the tyres and the road. However, as much friction as possible is needed when the brakes are applied. The tyres, too, must produce a large frictional force to accelerate and to slow down the car for cornering.

Sliding friction between two surfaces is constant whatever the relative speed of the two bodies. This is not so for a solid object passing through a fluid, where the resistance to motion increases with speed.

Simple machines

The simple machines considered here are devices in which a small force, the effort, can be made to produce a large force, the load.

The mechanical advantage of a machine is defined as the ratio of the load to the effort:

$$\text{mechanical advantage} = \frac{\text{load}}{\text{effort}}$$

Although a machine will produce a large force from a small one there cannot be a corresponding increase in energy. In fact there is a decrease in the work done on the load compared with the work done by the effort, because no machine is ever perfect and some energy is lost as heat caused by friction in the moving parts.

The effort will have to move a great distance to move the load a small distance. If the machine is moving at a steady rate the effort will move at a greater velocity than the load.

$$\text{velocity ratio} = \frac{\text{distance moved by the effort}}{\text{distance moved by the load in the same time}}$$

The **efficiency** of a machine indicates how good the machine is at doing its job. If the efficiency is low, a lot of energy is being wasted through friction in the moving parts of the machine.

$$\text{efficiency} = \frac{\text{work done on the load}}{\text{work done by the effort}}$$

Calculating the work, which is force × distance, it can be shown that:

$$\textbf{efficiency} = \frac{\text{load} \times \text{distance moved by load}}{\text{effort} \times \text{distance moved by effort}}$$

$$= \text{mechanical advantage} \times \frac{1}{\text{velocity ratio}}$$

Many lever devices produce large forces from small forces. The mechanical advantage will depend on the relative distances of the effort and the load from the fulcrum (figure 22).

Figure 22d shows how a crowbar works: the man can push up on the stone with a force of 9 times his own weight. But to move the load up 1 cm he will have to push down 9 cm.

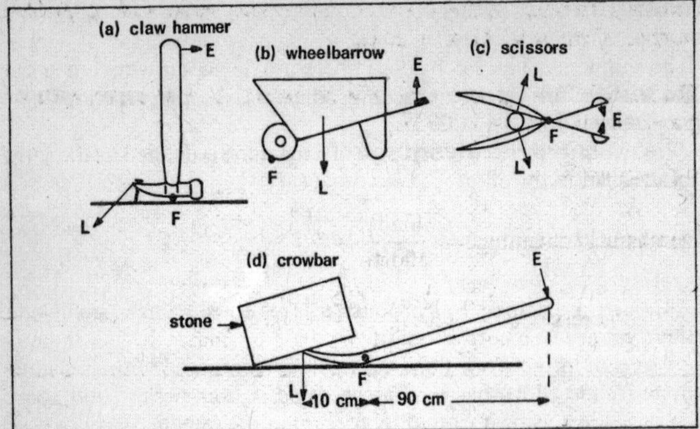

Figure 22. Lever devices

Another device for producing a large force from a small force is the pulley system. The pulleys will lose energy not only through friction but because the pulley block attached to the load has also to be lifted.

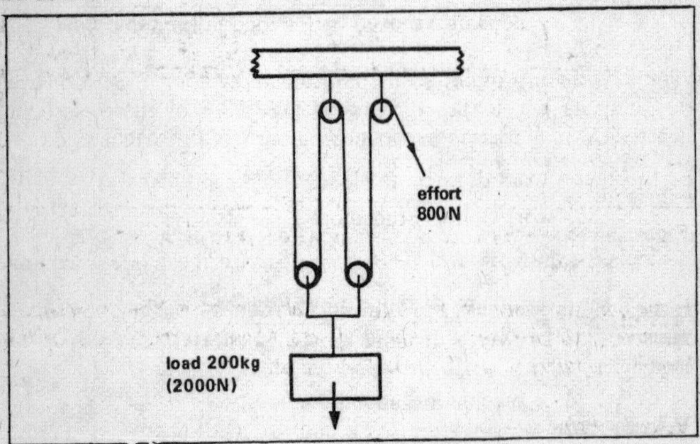

Figure 23. Pulley system

In the diagram of the pulley system (figure 23) the pulley blocks have not been drawn as they really are but spread out so that the ropes can be clearly seen. If the load is to be raised by 1 m, 1 m must be removed from each of the ropes holding the load. That makes a total of 4 m which must be taken up by the effort. The velocity ratio is therefore 4.

Similarly, in other pulley systems, the velocity ratio will equal the number of ropes holding the load.

The load is 200 kg and that will be pulled to the earth with a gravitational force of 2000 N.

Effort is 800 N.

$$\text{mechanical advantage} = \frac{2000\ \text{N}}{800\ \text{N}} = 2{\cdot}5$$

$$\text{efficiency} = \frac{2{\cdot}5}{4} = \frac{2{\cdot}5 \times 100\%}{4} = 62{\cdot}5\%$$

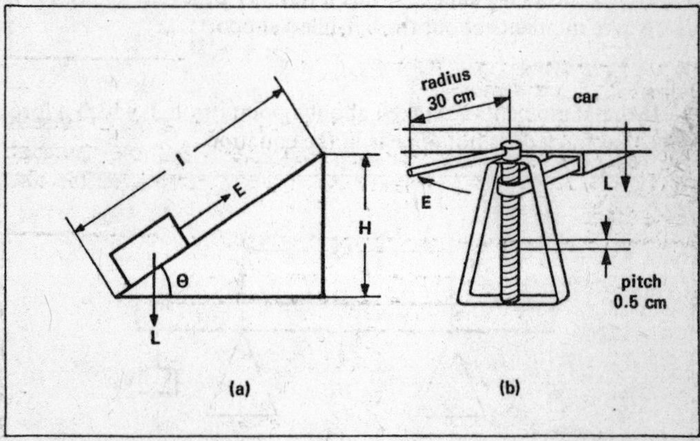

Figure 24. (a) Inclined plane; (b) screw-jack for lifting a motor vehicle

In the inclined plane (figure 24a) the distance moved by the effort is measured up the plane l in the diagram. The distance moved by the load is the vertical height H in the diagram.

$$\text{Velocity ratio} = \frac{l}{H} = \frac{I}{\sin\theta}$$

With a screw-jack (figure 24b) the velocity ratio is very high. The distance moved by the effort is the circumference of the circle described by the handle while the load moves only one screw pitch.

$$\text{velocity ratio} = \frac{2\pi \times R}{h} = \frac{2 \times 3{\cdot}14 \times 30}{0{\cdot}5} = 380$$

The efficiency is usually very low, at least lower than 50 per cent. Friction is high so that the screw does not run back if left.

Turning forces

The **moment of a force** about a point is the force × perpendicular distance of the force from the point.

Law of moments

If a body is in equilibrium then the sum of the clockwise moments about any point is equal to the sum of the anti-clockwise moments.

Figure 25 shows a plank that has a number of forces acting on it. If we take moments about the left-hand support:

$W_1 \times L_1 + W_2 \times L_3 = F_2 \times L_2$.

Note that if moments are taken about a point through which a force acts, that force does not appear in the equation.

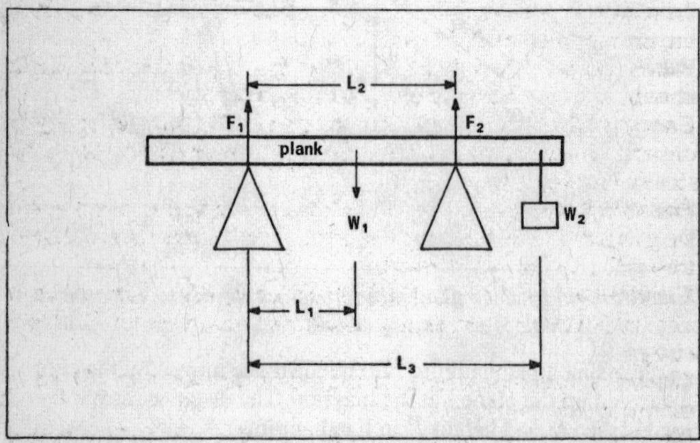

Figure 25. Law of moments

Key terms

Speed The distance moved in unit time. Units are metres/second (m/s).
Velocity Also the distance moved in unit time, but the direction of motion is also given or implied. For direction in a single line velocity can be positive or negative. Again units are metres/second (m/s).

Acceleration The rate of change of velocity. Units are metres/seconds^2 (m/s^2).
Force An influence which produces or attempts to produce motion or change of motion. A single force acting on a body will cause it to accelerate. Two or more balanced forces acting on a body will cause it to change shape. Units are newtons (N).
Weight The pull of gravity on a body. It varies from place to place. Units of force are newtons (N). Sometimes kilograms force (kg F) are used.
Mass A fundamental property of a body: the measure of the quantity of matter in it, to which its inertia is ascribed. Being a constant for that body, it does not alter from place to place, and is measured in terms of a defined unit, the kilogram (kg).
Inertia The property of matter by which it retains its state of rest (or of uniform motion in a straight line) so long as it is not acted upon by an external force.
Vectors Those quantities that have both magnitude and direction.
Momentum The mass of a body multiplied by its velocity. Its units are the same as for impulse, newtons × seconds (N.s).
Impulse Force multiplied by time of action of the force. Units are newtons × seconds (N.s).
Work Force × distance moved in the direction of the force. Units of work are newtons × metres, known as joules (J).
Energy A body or system contains energy if it has the ability to do work, i.e. can move a force through a distance. The units are the same as for work, joules (J).
Potential energy Energy arising from a body's position above the ground. A coiled spring will have potential energy because of the work it can do when released.
Kinetic energy Energy that is possessed by a body when it is in motion. It can do work as it comes to rest and gives up its kinetic energy.
Chemical energy Energy arising from the reaction of two or more chemicals with each other.
Heat energy The kinetic energy contained in the random motion of atoms or molecules.
Light energy Energy carried by light waves.
Electrical energy Energy produced by electricity.
Sound energy Energy produced by sound waves.
Nuclear energy Atomic energy, energy obtained from changes within the atomic nucleus, chiefly from nuclear fission or fusion.
Power The rate of doing work, work divided by time. Units are joules/seconds (J/s). This has a special name—watts (W).
Centripetal force The force that is required by a body to make

it move in a circle. It is always directed into the centre of the circle.
Friction The force built up between the two contacting surfaces when one body slides over another. It will tend to decrease their relative motion.
Simple machine One which changes the size and/or the position of action of a force.
Mechanical advantage The ratio of load to effort.
Velocity ratio That of the distance moved by the effort to the distance moved by the load.
Turning forces These are called moments. Their numerical value is calculated as the force multiplied by the perpendicular distance of the force from its point of action. Units are newtons × metres (N.m).

Examples

Velocity and acceleration

(1) If a car's brakes give it a uniform deceleration of 4 m/s^2, how far will it travel in coming to rest from 20 m/s?

The first problem when dealing with calculations of this type is to decide which of the equations of motion to use. It is helpful to write down the various quantities which are given. In this case $a = -4$ m/s^2, $u = 20$ m/s, $v = 0$ and we are asked to calculate distance, s.

In determining which equation of motion to use, it should be noticed that no mention is made of time, and the only equation in which this does not occur is $v^2 - u^2 = 2as$.

So $v^2 = u^2 + 2as$ gives $0 = 400 + 2.(-4)s$ from which s is found to be 50 m.

(2) A stone is thrown vertically upwards with an initial velocity of 15 m/s. Find the time for which it is in the air and the maximum height to which it rises. Take $g = 10$ m/s^2.

A number of points about vertical motion under gravity are illustrated in this example.
Firstly, the stone rises, stops momentarily, and then falls to earth again. This means that the motion has to be considered in two parts. For the rise, $u = 15$ m/s, $v = 0$, $a = -10$ m/s. The first task is to find the time for which the stone is in the air, so to begin with, the time, t, to rise should be found.

Using the equation $v = u + at$
$0 = 15 + (-10).t$
From which $t = 1{\cdot}5$ s.

To find the height to which the stone rises, use the equation $v^2 = u^2 + 2as$
Hence $(0)^2 = (15)^2 + 2.(-10).s$
from which $s = 225/20 = 11{\cdot}25$ m.

The time to fall can be calculated since the distance the stone falls is known to be the same as that which it rises, i.e. 11·25 m.
Using the equation $s = ut + \frac{1}{2}at^2$
$11{\cdot}25 = 0 + 0{\cdot}5.10.t^2$
So $t^2 = 11{\cdot}25/5 = 9/4$
and $t = \sqrt{9/4} = 3/2 = 1{\cdot}5$ s.
The total time of flight is therefore 3 s. It is useful to remember that **time to rise = time to fall**. With that knowledge the last part of the calculation could have been avoided.

Force, mass and acceleration

(1) What force is required to enable a car of mass 500 kg to reach a velocity of 24 m/s in 8 s, starting from rest?

(i) To calculate the acceleration, let $u = 0$, $v = 24$ m/s and $t = 8$ s.
Using the equation $v = u + at$
$24 = 0 + a.8$
Hence $a = 3$ m/s^2.

(ii) Use the equation $F = ma$ to find the force. $m = 500$ kg and $a = 3$ m/s^2.
Hence $F = 500 \times 3 = 1\,500$ N.

(2) A rifle bullet of mass 8 g, travelling at 800 m/s, buries itself 10 cm in the target. What is the average force exerted on the bullet during the impact?

The first point to notice here is that the units given are mixed and the information should be rewritten is S.I. units. Hence $m = 0{\cdot}008$ kg, $u = 800$ m/s, $v = 0$, $s = 0{\cdot}1$ m. Now the stages are as before.

(i) Calculate the acceleration. Using $v^2 = u^2 + 2as$,
$0 = (800)^2 + 2.a.(0.1)$
$a = -640\,000/0{\cdot}2 = -3\,200\,000$ m/s^2.
Notice that the answer is negative; this should have been expected because, obviously, the bullet is decelerating.

(ii) Use the equation $F = ma$ to calculate the force.
$F = 0{\cdot}008 \times -3\,200\,000 = -25\,600$ N.
It should be noted that this force is also negative. The point here is that the direction in which the force acts is opposite to that in which the bullet is travelling. This illustrates something of the vector nature of forces, because clearly a positive force would have caused the bullet to speed up rather than slow down. The example also serves to introduce **Newton's third law of motion** which states that to every action there is an equal and opposite reaction. In this case the action is the force exerted by the bullet on the target, and the reaction is the force exerted by the target on the bullet.

Conservation of momentum

A shell of mass 40 kg is fired from a gun of mass 8 000 kg. The muzzle velocity of the shell is 800 m/s. What is the recoil velocity of the gun?

The total momentum of gun and shell before the explosion is zero, since both are initially at rest. The law of conservation of momentum states that the total momentum after firing must also be zero,
$\therefore$ momentum of gun + momentum of shell = 0.
Since momentum in each case is mass × velocity,
$\therefore 8\,000 \times v + 40 \times 800 = 0$
Dividing through by 8 000,
$\therefore v + 4 = 0$
$\therefore v = -4$ m/s.
The minus sign indicates that the gun moves backwards, i.e. in the opposite direction to that of the shell. This is because shell and gun exert equal and opposite forces on each other.

Work, energy and power

(1) In tightening a nut, a man uses a spanner to exert a force of 25 N. This force acts at a distance of 14 cm from the centre of the nut. If it takes the man 4 seconds to turn the spanner through one complete revolution, find
(i) the work done in one complete turn;
(ii) the average power he exerts.

The work done = force × distance. The distance in this case is the circumference of a circle of radius 0·14 m, so
(i) work done = $25 \times 2 \times 22/7 \times 0{\cdot}14 = 22$ J;
(ii) power = work/time = $22/4 = 5{\cdot}5$ W.

(2) Discuss the energy changes which occur
(i) as a simple pendulum swings to and fro;
(ii) as a car starts from rest, is driven to the top of a hill, and again comes to rest.

The kinetic energy which the pendulum possesses as it swings through its lowest position is gradually converted to potential energy as the bob rises. At the top of the swing the energy is entirely potential, while at the lowest point it is entirely kinetic.

For full marks in such a question it should be mentioned that at intermediate points the sum of potential and kinetic energies is constant, assuming that there is no air resistance. In practice the amplitude of the swings gradually decreases, energy being lost in the form of heat owing to air resistance and friction at the point of suspension.

A considerable amount of detail is required to gain full marks in this type of question.

In the case of the car, the energy changes may be summarised as follows:
Chemical energy of fuel → heat energy (in the cylinder).
Heat energy → kinetic energy of pistons (in the cylinders).
K.E. of pistons → K.E. of car (crankshaft and transmission).
K.E. of car may change to P.E. if the car slows down as it goes up the hill. Otherwise the P.E. comes from the chemical energy of the extra fuel used up in the climb. K.E. → heat energy if the car is braked.

(3) Calculate the minimum power needed to raise 400 kg of water per minute from the bottom of a shaft 20 m deep. Why is the actual power required greater than this? (Take $g = 10\ \text{m/s}^2$.) Though not all the following stages may be required, they are given for ease of understanding.

Weight of water $= mg = 400 \times 10$ newtons
Work done on water $=$ force $\times$ distance $= 400 \times 10 \times 20$ joules
Time taken $= 1$ minute $= 60$ seconds, so

$$\text{power} = \frac{\text{work}}{\text{time}} = \frac{400 \times 10 \times 20}{60} = \frac{4\,000}{3} = 1\,333\ \text{W}$$

The actual power will be greater than this because
(i) there is friction in the moving parts of the pump;

(ii) the water emerging has some kinetic energy and the pump must have done additional work in order to give the water this energy.

Note The stages of the calculation have not been worked out, for this could lead to arithmetical errors, but have been left until the end. In many cases, as here, numbers will cancel.

The details given in the last part should also be noted. Mere reference to friction, with no explanation, would only gain about a quarter of the possible marks.

Machines
A pulley system has a V.R. of 8. The lower block weighs 6 N and a load of weight 24 N is attached to it. An effort of 6 N is needed to lift the load. Calculate the efficiency of the system and the work done by the effort in raising the load a distance of 2 m.
Why is the efficiency of a pulley system never 100 per cent?

Calculation:
M.A. = load/effort = 24/6 = 4
efficiency = M.A./V.R. = 4/8 or 50%
For the load to be raised through 2 m, the effort must move through $2 \times 8 = 16$ m.
∴ Work done by effort = force × distance = $6 \times 16 = 96$ J.

There are two main reasons why the efficiency of a machine is never 100 per cent. Firstly there is always some friction in the moving parts, and extra work must be done in order to overcome this. Secondly, work must be done to move the machine itself. This point is illustrated in the example, where the lower block has a significant weight.

Chapter 3
Heat

Thermometers

A thermometer measures **temperature**, or the **degree of hotness**. Any substance which possesses one or more properties which vary with the degree of hotness can be used to measure temperature. The most usual such substance is mercury, which has the property of expanding when its temperature rises. The expansion of mercury in a glass tube on heating illustrates conveniently the usual range of temperatures used in a school laboratory, but does not cover the full range required for industry and research work.

Other properties which may be used to measure temperature are the **electrical resistance** of platinum and the **thermoelectric current** generated at the junction of two metals when heated.

Celsius scale of temperature

The lower fixed point is the temperature at which water freezes (figure 26a). **The upper fixed point is the temperature at which water boils** (figure 26b) under standard atmospheric pressure, that is 760 mm of mercury. It was **Celsius**

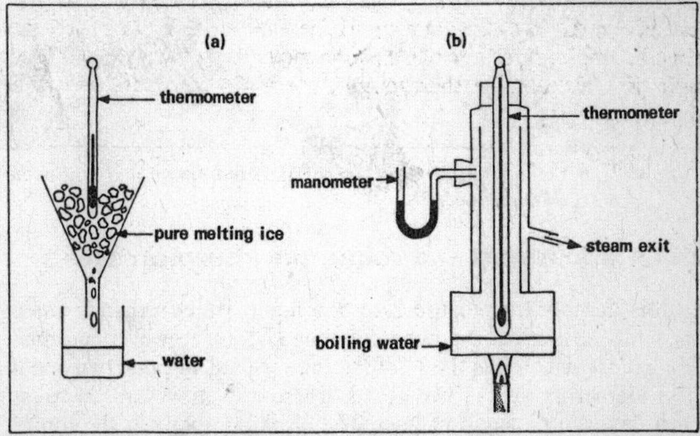

Figure 26. Marking (a) the lower and (b) the upper fixed points of the Celsius temperature scale

who first suggested that the interval between the two points be divided into **one hundred parts**. The units of temperature on this scale, now called the Celsius scale, are degrees Celsius (°C).

Clinical thermometers

Clinical thermometers are specially designed for finding the temperature of the human body (figure 27a). They have a limited range, from **35°C to 43°C**, and a very narrow capillary which enables them to be easily read to an accuracy of 0·1°. Another adaptation is the **constriction** just above the bulb up through which the expanding mercury pushes when the temperature rises. However, when the temperature drops and the mercury contracts, the mercury in the tube does not drop down into the bulb. Thus, the **maximum temperature** registered by the mercury can be read at leisure, and the mercury later forced back into the bulb by shaking.

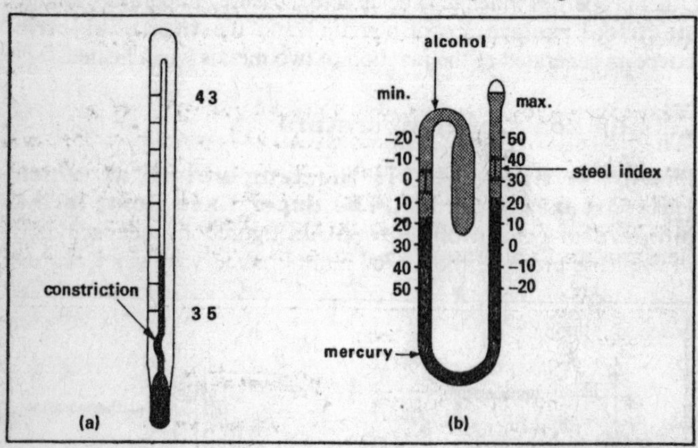

Figure 27. (a) Clinical thermometer; (b) Six's maximum and minimum thermometer

Six's maximum and minimum thermometer

In this thermometer (figure 27b), the liquid which expands when the temperature rises is not mercury but alcohol. However, a thread of mercury which fills the U of the tube is used to push two small iron indicators up the two scales. These indicators can be re-set each day using a small magnet. The alcohol vapour in the top of the right-hand tube acts as a sort of spring which pushes the mercury back as the temperature drops.

Absolute scale of temperature

This, known as the Kelvin scale, will be explained more fully on page 58. To calculate the Kelvin temperature from the Celsius temperature **add 273**. Thus 0°C is 273 K (Kelvin), 100°C is 373 K and 27°C is 300 K.

Expansion of solids on heating

Though the effect of rising temperature on solids is small and difficult to detect in laboratory experiments, engineers must allow for expansion and contraction in all design projects likely to be affected by temperature changes. Examples include railway lines, bridges and concrete road surfaces. The differences in temperature between a hot summer day and a cold winter night could cause expansion and contraction to the extent of several centimetres per twenty-five metres of rail, bridge structure or road surface.

Thermal expansion shown in a bi-metallic strip

This experiment (figure 28a) demonstrates how rods of the same length made of different metals expand to different lengths on heating. For example, brass expands 50 per cent more than iron. Therefore if rods of iron and brass are fixed together at room temperature then heated, the heated strip will curve so that the

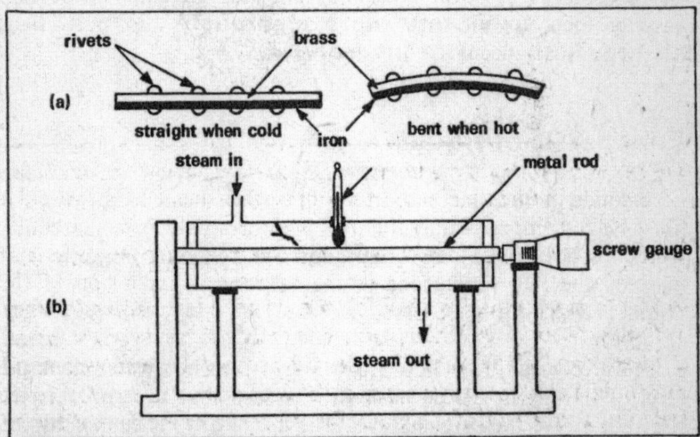

Figure 28. Thermal expansion: (a) bi-metallic strip; (b) apparatus for finding the coefficient of linear expansion

brass is on the outside. Such devices are found in **thermostats** and **motor-vehicle flasher units**.

Coefficient of linear expansion

The coefficient of linear expansion is **the fractional change in length for one degree rise in temperature**. Its value can be determined using the apparatus shown in figure 28b.

To find the coefficient, α, measure the original length of the bar, L, the change in temperature, θ, and the change in length, l.

The length, L, need only be measured with a metre rule to the nearest millimetre, whereas the change in length, l, should be measured with a micrometer to the nearest 0·01 mm.

$\alpha = \dfrac{l}{L \times \theta}$ Since the units of l and L are the same they cancel so the units of α are /°C.

Typical values of α are: iron 0·000 012/°C;
brass 0·000 019/°C.

Kinetic theory and the expansion of solids

The kinetic energy of the molecules of a solid increases on heating causing them to **vibrate more vigorously** and push their neighbouring molecules a little further away.

Expansion of liquids

Like solids, liquids increase in volume when heated. However, it must be remembered that a liquid is always held in a container, which will be heated along with the liquid and expand. The difference between the two expansions or the **apparent** expansion of the liquid, is what will be observed.

If a flask containing a liquid is plunged into hot water then the level of the liquid in the flask will be seen first to drop then to rise well above the original level. The initial drop is caused by the expansion of the glass that has been heated first. The liquid, when heated, will expand much more than the solid.

Change in density in relation to expansion of liquids

For a fixed mass of liquid the volume increases.

$$\text{density} = \frac{\text{mass}}{\text{volume}}$$ Therefore the density decreases.

It has already been demonstrated that a less dense substance floats on top of a more dense substance, therefore the heated liquid will rise. (See page 61, convection currents.)

The unusual expansion of water

If water at 0°C is heated it **contracts, becoming more dense, until it reaches 4°C**. Above this temperature it expands and decreases in density like other liquids (figure 29a).

As a result, when lakes and ponds freeze in winter, the most dense water at 4°C remains at the bottom while the less dense water at the top is colder and will freeze (figure 29b).

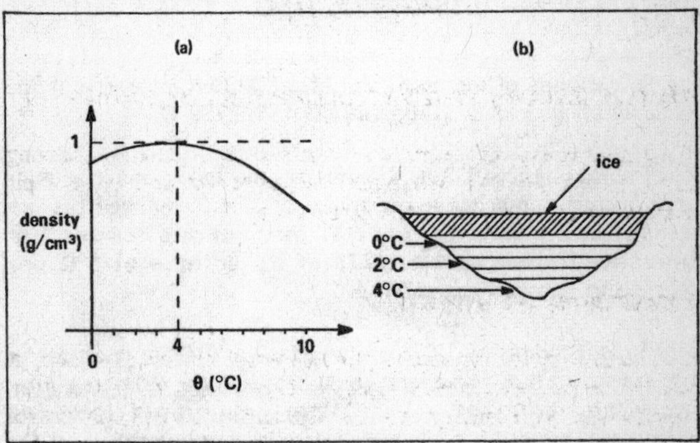

Figure 29. (a) Variation of water density with temperature; (b) temperatures in ice-covered pond

Expansion of a gas

If a gas is heated at a fixed pressure it expands. Gases expand to a far greater extent than either liquids or solids.

As in liquids and solids, as the volume increases the density decreases and convection currents can be set up. Hot-air ballooning, which has become so popular over the last few years, demonstrates this principle.

Coefficient of expansion of a gas at constant pressure

This is the fraction of its volume at 0°C by which it expands for each degree rise in temperature.

In the experiment illustrated below to find the coefficient of the expansion of a column of gas, the length of the column of air trapped in the tube by the mercury thread may be considered proportional to the volume (figure 30a).

A graph is drawn of length against temperature and the coefficient, α, can be calculated from the values L_0 and L_{100} obtained from the graph (figure 30b):

$$\alpha = \frac{L_{100} - L_0}{L_0 \times 100}$$

Again, the units of the coefficient are **/°C**. The value of α is the same for all gases and is **0·003 66/°C** or **1/273/°C**.

The French physicist J. A. C. Charles was the first to undertake this experiment using accurate apparatus, and he formulated the law stating that **the volume of a fixed mass of gas at constant pressure increases by 1/273 of its volume at 0°C per °C rise in temperature.**

If the gas increases by 1/273 of its volume at 0°C for each degree rise in temperature, the reverse could happen if the gas were cooled below 0°C. **Thus at −273°C the gas would have no volume.** This can also be seen from the graph opposite: if the straight line were continued downwards it would cross the temperature axis where L is zero at −273°C.

In practice most gases will change into a liquid long before this temperature is reached. In theory a **'perfect gas'** is one that will have zero volume at this temperature because the **molecules are no longer moving.**

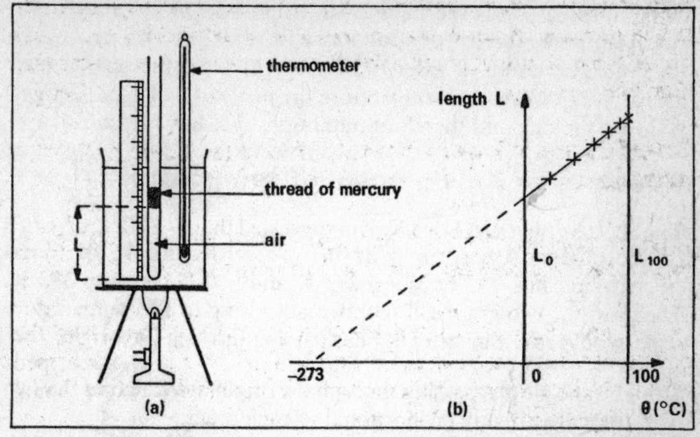

Figure 30. (a) Charles' law apparatus; (b) graph showing variation of length of gas column with temperature

Variation of pressure of a gas kept at constant volume

When a gas is heated it usually expands and also increases in pressure as well. In the previous experiment the pressure was deliberately kept constant. However, if a fixed volume of gas is heated (figure 31a), the pressure will increase.

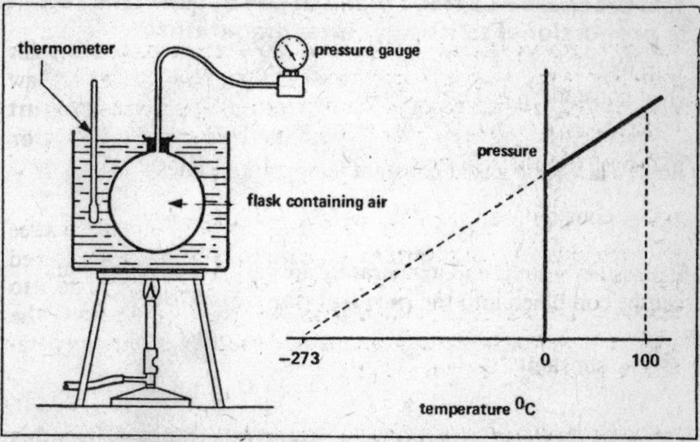

Figure 31. (a) Apparatus and (b) graph illustrating the pressure law

The resulting graph (figure 31b) will be very similar to the one shown in figure 30b but this time shows the pressure changing. The slope of the graph is the same. **The increase in pressure for each degree rise in temperature is 1/273 of the pressure at 0°C**. The graph, if extended back, will cross the temperature axis at −273°C. At this temperature the molecules of a 'perfect gas' will stop moving and therefore cannot produce any pressure.

Absolute or Kelvin scale of temperature

The Scottish physicist Lord Kelvin suggested that **−273°C should be considered the zero of his absolute scale of temperature**. Thus 0°C becomes 273 K and 100°C becomes 373 K. The previous two graphs, if redrawn according to this temperature scale, would result in straight lines passing through the origin.

Straight-line graphs passing through the origin indicate that the two quantities are directly proportional to each other.

Charles's law can be restated:

The volume of a fixed mass of gas at constant pressure is proportional to its absolute temperature.

$$\frac{V}{T} = \text{constant}$$

The pressure law can be stated similarly:

The pressure of a fixed mass of gas at constant volume is proportional to its absolute temperature.

$$\frac{p}{T} = \text{constant}$$

Boyle's law for a gas at constant temperature states that;

$$p.V = \text{constant}$$

If pressure, volume and temperature are varied these three equations can be combined into the **general gas equation**:

$$\frac{p.V}{T} = \text{constant}$$

Often in chemical experiments a gas is collected and its volume measured at the pressure and temperature of the laboratory. By referring to a list of densities of gases at **standard temperature and pressure (s.t.p.,** i.e. at **0°C and 760 mm of mercury)**, it is possible, using the gas equation, to find the volume of a gas at s.t.p. and thus calculate its mass.

e.g. Laboratory values: $p = 750$ mmHg, $V = 100\ \text{cm}^3$, $T = 17°\text{C}, 290\ \text{K}$; s.t.p.: $p = 760$ mmHg, $V = ?$, $T = 0°\text{C}, 273\ \text{K}$

$$\frac{750 \times 100}{290} = \frac{760 \times V}{273}$$

$$V = 92{\cdot}9\ \text{cm}^3.$$

Example 1

A bottle contains air at 27°C. The atmospheric pressure is $10^5\ \text{N/m}^2$. A bung is placed in the bottle which is then heated to 387°C, whereupon the bung of the bottle blows out. Calculate the pressure of the air needed to blow out the bung. Any expansion of the bottle may be neglected, but explain why such an approximation is justified.

The calculation is a straightforward application of the relationship between the pressure of a fixed mass of gas and its temperature, i.e.

$\frac{p_1}{T_1} = \frac{p_2}{T_2}$ but remember to change the temperature to degrees Kelvin.

$$\therefore \quad \frac{10^5}{(27+273)} = \frac{p_2}{(387+273)}$$

$$\therefore \quad p_2 = \frac{10^5 \times 660}{300} = 220\,000\ \text{N/m}^2.$$

In general the expansion of solids is very much less than the expansion of gases. It is usual in this type of calculation to ignore any change in volume of the vessel in which a gas is contained. The explanation at the end of the question should mention that the expansion of glass may be ignored compared with the expansion, or pressure increase, of the air.

Example 2

A certain mass of gas is heated at constant pressure. The volume increases by 10 cm³ when the temperature of the gas is raised from 0°C to 20°C. Find (i) the volume of the gas at 0°C; (ii) its volume at 100°C.

The problem may be solved by an application of Charles' law, but is slightly more complicated than previous ones, since neither the initial nor the final volumes of the gas are given directly. Instead the increase in volume for the stated rise in temperature is given. If V_0 represents the volume of 0°C, then the volume at 20°C is V_0+10.

Applying Charles' law:

$$\frac{V_0}{273} = \frac{V_0+10}{293}$$

$$\therefore \quad 293V_0 = 273V_0 + 2730$$

$$\therefore \quad 20V_0 = 2730, \quad \text{from which} \quad V_0 = 136{\cdot}5 \text{ cm}^3$$

For the second part:

$$\frac{136{\cdot}5}{273} = \frac{V_{100}}{373}$$

$\therefore V_{100} = 186{\cdot}5 \text{ cm}^3$ (Note the cancelling: the left-hand side of the equation reduces to $\frac{1}{2}$.)

Problems on the gas laws are usually fairly simple, but the importance of converting temperatures to degrees **Kelvin** cannot be emphasised too strongly. Failure to do this renders the whole calculation valueless and would gain a zero mark for the question.

Example 3

A fixed mass of air at a temperature of 15°C is enclosed in a wide capillary tube by an index of sulphuric acid. The pressure resulting from the short length of the index of sulphuric acid may be neglected. The atmospheric pressure is 0·76 m of mercury. The air column is 0·12 m long. Find its length if the temperature is raised to 87°C and atmospheric pressure falls to 0·75 m of mercury.

Applying the general gas equation:

$$\frac{0{\cdot}76 \times 0{\cdot}12}{288} = \frac{0{\cdot}75 \times l}{360}$$

Note that the lengths are used to represent the volumes of the air, and this is justified if the cross-sectional area of the tube is constant.

This equation gives $l = \dfrac{0{\cdot}76 \times 0{\cdot}12 \times 360}{0{\cdot}75 \times 288} = 0{\cdot}152 \text{ m}$

Transmission of heat energy

Conduction

If a metal rod is held at one end and the other end put into a bunsen flame, before long the rod will become too hot to hold. Heat has been **conducted** along the rod. However, a similar-sized glass rod will not show this effect. The end in the flame will become very hot and may melt while the end being held remains cold.

Metals, which are **good conductors of heat**, are also good conductors of electricity. The electric current is carried by the **free electrons** (see page 128) **in the metal**. Physicists think that these free electrons are also responsible for carrying the heat along the metal but there is no experiment that can be done in a school laboratory to demonstrate this directly. **Solid non-metals**, **liquids** and **gases** are **very poor conductors of heat**.

Convection

Convection currents can be demonstrated to move through **liquids** and **gases** by the apparatus shown in figure 32. In (a) the heated water at the bottom of the beaker expands, becomes less dense, thus causing it to rise. Colder water close to the sides of the flask descends to take its place. In (b) air above the candle flame is heated, becomes less dense, and rises up the funnel. Colder air is drawn down the other funnel to take its place. In both cases, heat is carried by the upward movement of the water or air in bulk.

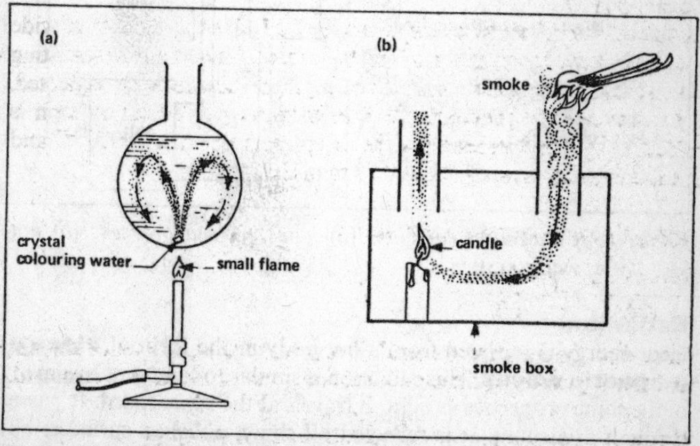

Figure 32. Convection currents: (a) in water; (b) in air

Practical examples of convection currents

On a worldwide scale convection currents cause air to move and therefore affect climate. This can be illustrated on a smaller scale by land and sea breezes, as indicated in figure 33a. In each case, air in the relatively warmer area rises to be replaced by air from the relatively cooler area moving across the surface of the land, or sea, causing a breeze.

Domestically, convection currents are used to maintain and assist the circulation of water in a central heating or hot-water system. In both cases water is heated in the boiler, which is at a lower level than the rest of the system. The heated water then rises by convection, either to circulate through a central heating system, often assisted by pumps, and eventually descending again to the boiler on losing its heat; or to be stored in a well-insulated tank to supply hot water when needed. In the latter case the boiler is replenished by cold water from a supply tank high in the building, or by water in the hot tank which has cooled and descends again to the boiler (figure 33b).

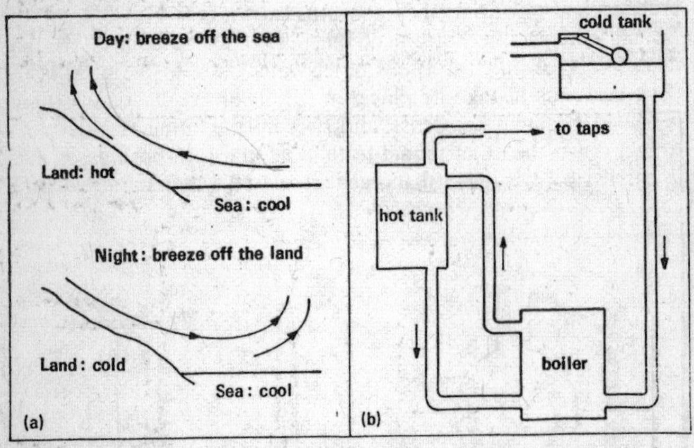

Figure 33. Convection currents: (a) land and sea breezes; (b) hot-water system

Radiation

Heat energy is radiated from a hot body in the form of **electro-magnetic waves**. This radiation is similar to light and has many of the same properties as light. It travels at the same speed. It passes through a vacuum. It is reflected off shiny, polished surfaces, but absorbed by dull, black surfaces.

Dull, dark coloured surfaces are good absorbers of radiation when cool, and good emitters of radiation when hot. Polished light-coloured surfaces, on the other hand, are poor absorbers when cool and poor emitters when hot.

This has various practical applications such as the wearing of white clothing in hot climates and the use of polished silver tea-pots.

The heat energy as well as the light energy from the sun comes to us in the form of radiation. This radiation consists of **electro-magnetic waves** of different **wavelengths**. The range of wavelengths which transmit heat energy are known as **infra-red**.

The greenhouse effect

Only short-wavelength infra-red waves, along with light, will pass through glass. Therefore heat and light energy from the sun passes through the glass and is absorbed by the objects in the greenhouse. These hot objects emit a radiation of a much longer wavelength, that cannot pass out through the glass. The energy of this longer-wavelength radiation is thus trapped inside the greenhouse as heat energy, and the temperature inside the greenhouse rises (figure 34a).

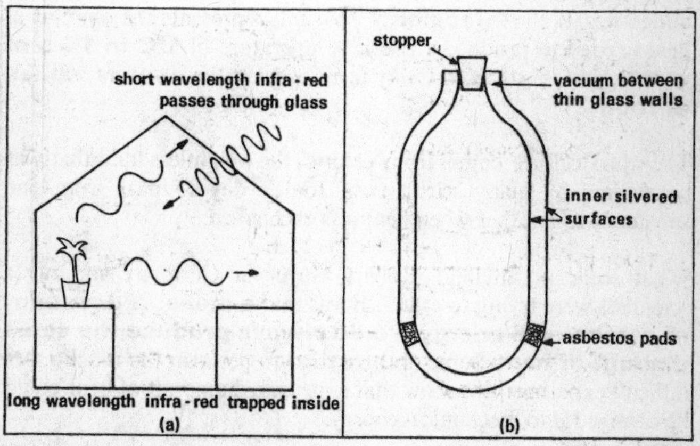

Figure 34. Heat transfer: (*a*) *greenhouse effect;* (*b*) *vacuum flask*

Vacuum flasks

Vacuum flasks are a good example of **prevention** of all three types of heat transfer (figure 34b).

Only a very little heat can be transferred by conduction along the thin glass.

No convection is possible because the space between the two glass layers is empty (a vacuum).

Radiation is minimised by silvering the two glass layers. The hot silver surface emits a little radiation and the cold silver surface will reflect back what radiation is emitted.

Note Vacuum flasks keep hot substances hot and cold substances cold: they **prevent transfer of heat** in either direction.

Heat energy

The idea that **heat** is a **form of energy** which can be measured in terms of our normal energy unit, the joule, is relatively modern.

The joule was firmly established during the nineteenth century after careful experimental evidence had been produced by a number of scientists, the most important of whom was **James Joule**.

Heat used to be measured in a variety of units. The easiest to understand is the **kilocalorie**. This unit represents the amount of heat needed to produce a rise in temperature of **1°C in 1 kg of water**, and is still used today in measuring the **energy values of foods**.

The word 'calorie' comes from 'caloric', the weightless fluid that was considered by many scientists in Joule's day to pass from one substance to another when heat was exchanged.

What Joule in England, Robert Mayer in Germany and other scientists were trying to establish was that **a measured amount of mechanical energy would always produce the same amount of heat**. Some of them tried to perform the much more difficult experiment to show that a measured amount of heat could be changed into mechanical energy.

Joule's paddle-wheel experiment

In the experiment shown in figure 35, the lead weights fall through a measured distance, using up a measured amount of potential

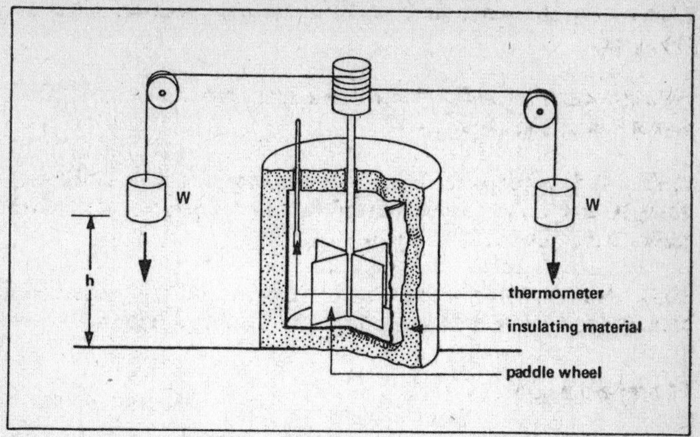

Figure 35. Joule's paddle-wheel experiment

energy. The churning action of the paddle wheels causes the temperature of the water and its container to rise.

From a previous experiment which involved mixing hot and cold water, Joule knew the heat capacity (see page 67) of the water, paddle wheel and container. By multiplying the heat capacity by the temperature rise he calculated the heat gained in the old British units that were equivalent to the kilocalorie. **Energy gained by system = heat capacity × temperature rise.**

He performed many experiments of this type, making careful allowances for heat lost to the air, for friction at the pulleys and for the kinetic energy of the weights when they hit the floor. Each experiment produced the same result: **4200 joules of energy disappeared and 1 kilocalorie was produced.**

In figure 36 three further experiments in converting various forms of energy to heat are illustrated.

(a) In 1798 Rumford carried out qualitative work on the heat generated while boring cannons. He discovered that blunter tools cut less efficiently but produced more heat.

(b) In 1861 Hirn crushed a lead block and equated the kinetic energy of the moving iron with the heat energy produced in the lead.

(c) Lead shot falling in a cardboard tube should gain a quantity of heat equivalent to the potential energy lost by the lead in falling the

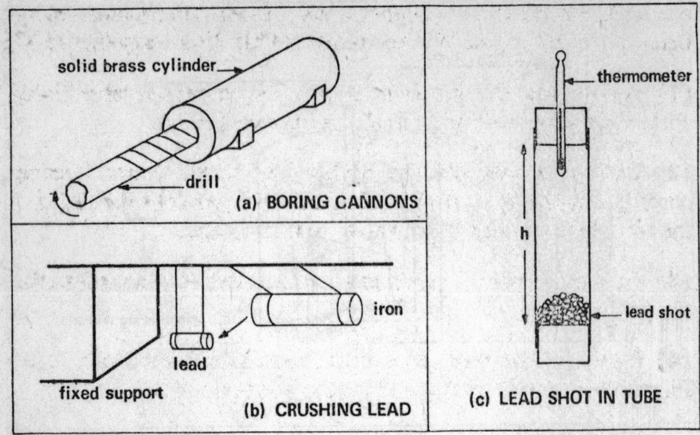

Figure 36. Experiments in converting various forms of energy to heat

length of the tube. If the tube is inverted many times in quick succession, a measurable temperature rise should be produced. In practice, the experiment does not give very accurate results.

The energy values of fuels and foods are found by burning them and noting the temperature rise in a vessel containing water surrounding the burning substance. This vessel is called a **bomb calorimeter**, and the experiment is carried out under carefully controlled conditions to ensure that burning is complete and that as little heat as possible is lost.

Example A bomb calorimeter and its contained water has a heat capacity of 5 000 J/°C. 0·01 kg of a food is burnt in it and its temperature rises by 5°C.

Energy given out by 0·01 kg of food = $5 \times 5\,000$ J = 25 000 J.

Therefore energy given out by 1 kg of food = 2 500 000 J or $2{\cdot}5 \times 10^6$ J.

Other energy values:
butter: $3{\cdot}3 \times 10^7$ J/kg;
beef: $0{\cdot}59 \times 10^7$ J/kg;
bread: $1{\cdot}01 \times 10^7$ J/kg.

The four-stroke internal combustion engine

This provides us with an example of heat energy being converted into mechanical energy. The source of heat in this case is the chemical reaction between petrol and oxygen, mechanical energy being produced in the form of rotational kinetic energy (figure 37).

(1) Induction stroke Inlet valve open, exhausts valve closed. Piston goes down, drawing in the petrol-air mixture.

(2) Compression stroke Both valves closed. Piston goes up, compressing the mixture. **Spark** ignites the gas with the piston at the top of the cylinder. Both valves remain closed.

(3) Power stroke The hot gas increases in pressure and pushes the piston down.

(4) Exhaust stroke Inlet valve closed, exhaust valve open. Piston goes up and pushes burnt gas out of exhaust.

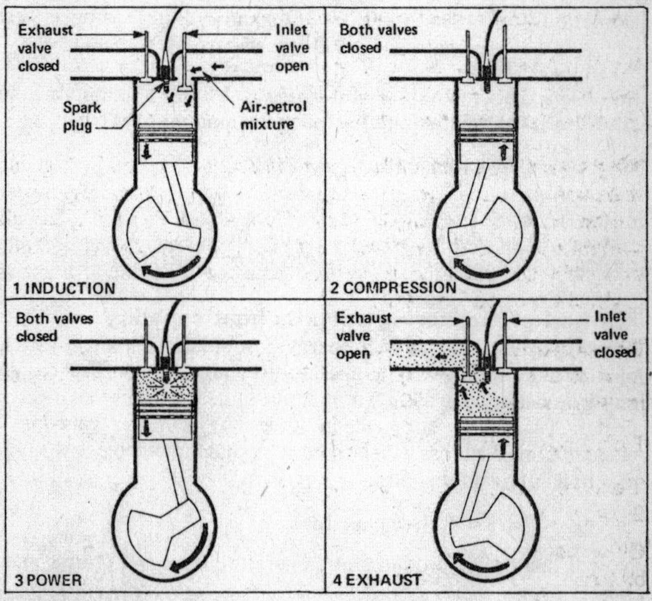

Figure 37. The four strokes of the internal combustion engine

Heat capacity

The heat capacity of a body is defined as the energy required to raise the temperature of the body by 1°C.

The value of the heat capacity for any body will depend on the material of which it is made, and its mass. The unit of heat capacity is J/°C. The concept of heat capacity can be applied to two or more substances in thermal contact, for example a calorimeter containing a liquid. (See example on page 66.)

Specific heat capacity

The specific heat capacity of a substance is defined as the energy required to raise the temperature of 1 kg of the substance by 1°C.

The specific heat capacity of water is **4 200 J/kg°C.**

Other specific heat capacities in the same units:

aluminium: 900	lead: 130	mercury: 140	copper: 400
ice: 2 100	paraffin: 1 200	seawater: 3 900	sand: 800.

So raising the temperature of a block of 4 kg of aluminium from 100°C to 160°C will require the following amount of energy:

$4 \times 900 \times (160 - 100)$ joules.

The general rule for calculating the energy required to raise a mass, m kg of a substance with a specific heat capacity c J/kg °C, through a temperature change of θ°C is: $\boldsymbol{m \times c \times \theta}$ **joules**.

Method of measuring specific heat capacity

A measured quantity of heat energy is supplied to a known mass of a substance, and the temperature change noted. The easiest method of supplying a known quantity of heat is to use an electric heater. The energy supplied, in joules, by a heater carrying a current, I amps, across which there is a potential difference, V volts, in a time interval t seconds is:

energy $= I \times V \times t$ (see page 136).

Now if this heater produces a temperature change θ°C in a mass m kg:

$$I \times V \times t = m \times c \times \theta$$

or $$c = \frac{I \times V \times t}{m \times \theta}$$

Allowances have to be made for the heat given to the heater itself and to the container if the substance is a liquid, also for the loss of heat to the air and surroundings.

Latent heat

To change a solid into a liquid energy must be supplied. This energy does not increase the temperature of the solid but changes its state to a liquid. It is called the **latent heat of fusion**. ('Latent' means 'hidden' and is used in this instance because the effect is 'hidden' from a thermometer.) Similarly, energy must be supplied to change a liquid into a gas. Again there is no change in temperature, only a change in state. This energy is called the **latent heat of vaporisation**.

Cooling curve for naphthalene

A little naphthalene (moth balls) is heated in boiling water to 100°C, at which temperature it will have changed to a liquid state. If a temperature/time graph is plotted as it cools, the graph will show a very marked horizontal line (figure 38). During this time the naphthalene will change back from liquid to solid but there will be no change in temperature. However, the naphthalene is still giving out heat, since heat will always pass from a hotter body, in this case the naphthalene, to a colder body in contact with it, in this case the air.

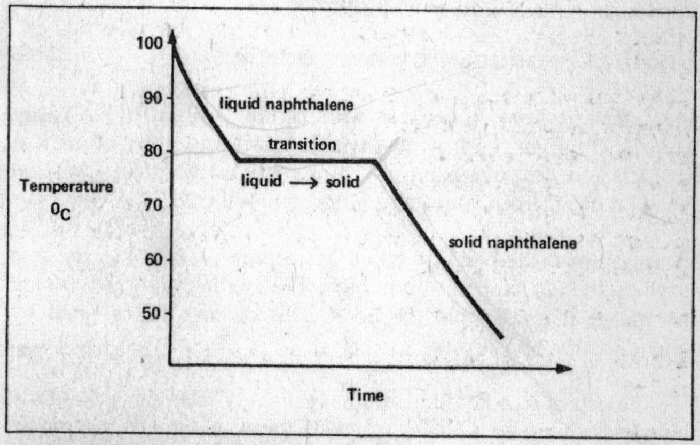

Figure 38. Cooling curve of naphthalene

Latent heat and kinetic theory

Latent heat of fusion is the energy required by the molecules to break the bonds which hold them in their fixed place within the crystal structure of the solid. In the liquid state, the molecules are almost as closely packed (when solids melt there is only a slight volume change, see page 9), but are free to move around within the body of the liquid. When a liquid solidifies, this energy is lost by the atoms and given off as latent heat.

Latent heat of vaporisation is the energy needed to overcome the close-range forces of attraction between the molecules which hold them in the body of a liquid.

Specific latent heat

The specific latent heat of fusion of a substance is the energy required to convert 1 kg of the substance from the solid to the liquid state without change of temperature.

Specific latent heat of ice = 336 000 J/kg.

The specific latent heat of vaporisation of a substance is the energy required to convert 1 kg of the substance from the liquid to the gaseous state without change of temperature.

Specific latent heat of steam = 2 260 000 J/kg.

Cooling produced by evaporation

This effect is very noticeable with certain **volatile** (or easily vaporised) liquids, such as methylated spirit and ether. If such a liquid is put on your hand, it will quickly evaporate taking the energy it needs from your hand, which will feel cold.

Perspiration is the body's method of keeping cool during exercise or when the air temperature is high. The sweat evaporates taking the energy it needs from the body. This cooling is explained by kinetic theory, as follows.

The molecules in a liquid, as we have already seen, are constantly moving about. The average speed of their movement is directly connected with the temperature of the liquid; the higher the average speed the higher the temperature. But the molecules are not all moving at this average speed: some are moving faster, some more slowly.

It is the faster-moving molecules that will have enough energy to escape through the surface and evaporate. If the liquid loses the faster-moving molecules the average speed of the remaining molecules will decrease, therefore the temperature will be lower (see also pages 10–11).

Action of a refrigerator

The cooling of a liquid during evaporation is the principle on which a domestic refrigerator works. A volatile liquid is allowed to evaporate inside a coil surrounding the freezing compartment, absorbing latent heat and cooling the compartment. The vapour is immediately removed by a pump to the condenser, which is situated outside the cabinet. Here the vapour is compressed and changed back into a liquid, giving out latent heat. The liquid is then pumped back to the coil around the freezer, and the cycle is repeated (figure 39a).

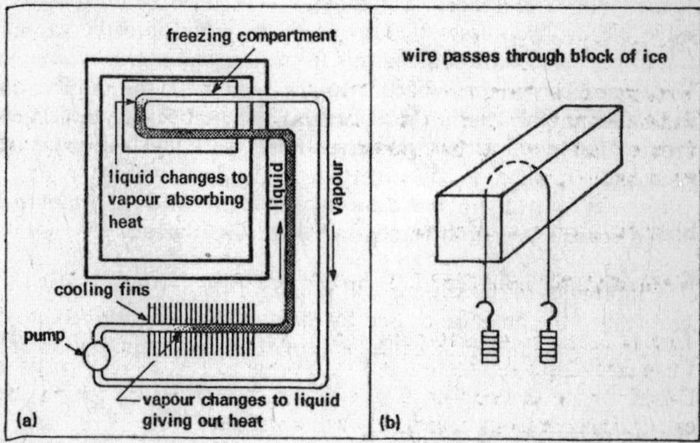

Figure 39. (a) Action of a refrigerator; (b) regelation

Changes of volume and the effect of pressure on melting temperatures

Regelation

Most solids increase in volume slightly when they change into a liquid. Any pressure on the solid will make it more difficult to change and therefore the temperature of melting will rise.

With a few substances, noticeably water, the reverse is the case. Ice, when it changes into water, decreases in volume. Any external pressure will assist the process so it will take place at a lower temperature. A popular demonstration of this effect is the passing of a piece of wire through a block of ice, the ice remaining in one piece. The wire increases the pressure on the ice immediately below it, that ice melts and the wire passes through. The water just above the wire is at normal pressure so will freeze again, a process known as **regelation** (figure 39b).

Saturated and unsaturated vapours

The fast-moving molecules in a liquid, as we have seen, may have enough energy to escape through the surface into the space above forming a **vapour**. This can happen at any temperature but will take place more quickly at higher temperatures because there will be more molecules with the necessary energy to escape.

If the liquid is in a closed container, a corked bottle for instance, molecules will continue to escape from the liquid into the vapour but at the same time some atoms from the vapour will strike the liquid surface and pass back into the liquid. As the number of molecules in the vapour increases, so the number of molecules striking the liquid surface and re-entering the liquid will increase. Eventually a state of **dynamic equilibrium** will be reached **when equal numbers are escaping from and re-entering the liquid**. The vapour is then said to be **saturated**.

The molecules of the vapour will bombard the walls of the vessel and add to the pressure caused by the air. The pressure produced by saturated vapour is called **saturated vapour pressure (s.v.p.)**.

Measurement of s.v.p.

If a small quantity of liquid is introduced to the top of the mercury column of a barometer, its vapour will fill the previously evacuated space. If enough liquid is present to allow the vapour to reach saturation, the pressure on the mercury column will be the s.v.p. of the liquid.

The difference between the heights of the mercury with and without the saturated vapour in the tube gives the s.v.p. in mmHg (figure 40a).

By enclosing the top of the barometer in a water bath, or other means of heating, the variation of s.v.p. with temperature can be

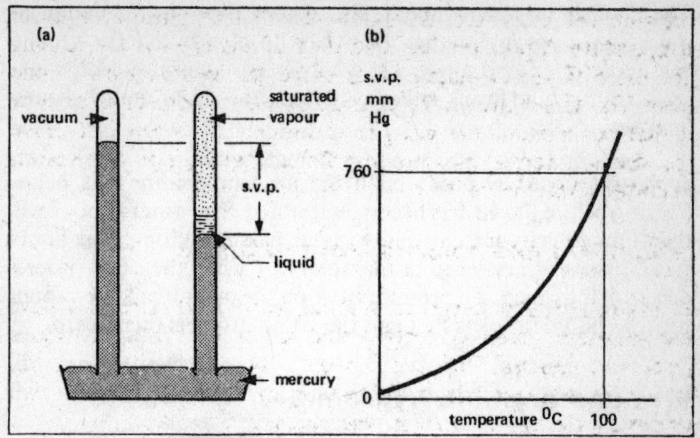

Figure 40. (a) Measuring s.v.p.; (b) variation of s.v.p. of water vapour with temperature

found. Notice that with **water the s.v.p. at 100°C is 760 mm.Hg**, i.e. normal atmospheric pressure (figure 40b).

The s.v.p. of a substance increases with temperature. The molecules in the liquid are, on average, moving faster so more escape into the vapour. Also the atoms in the vapour will be moving faster and therefore hitting the wall of the vessel harder.

Boiling occurs when the s.v.p. is equal to the external pressure on the liquid surface. Vapour can be produced in the liquid and bubbles appear. By **changing the external pressure we can change the boiling temperature of liquids**. Pressure cooking is an example of an increase in pressure producing a higher boiling temperature, and therefore a shorter cooking time. On the other hand, if the external pressure is reduced, either naturally, as at the top of a mountain, or artificially with a pump, the boiling temperature can be much reduced.

Water vapour in the atmosphere

The air blowing over the sea collects water vapour and may become saturated. If that air rises or for some other reason becomes colder, the air will be unable to hold so much water vapour and some will have to change back into liquid. **This liquid is often in the form of very small drops that fall through the air very slowly, in the form of clouds.**

The cooling of water vapour also causes dew. Warm damp air quickly cools at night when there are no clouds and the ground also radiates its heat away. The air becomes saturated with water vapour and liquid forms, either as **mist** in the air, or on the ground as **dew**.

A white frost is produced if the air temperature falls below freezing before the air has become saturated. The water vapour will then turn directly into a solid without passing through the liquid state. **Snow** is produced in the same way when the same process occurs in the clouds. To produce **hail**, on the other hand, the vapour turns to liquid droplets first and the liquid droplets then freeze.

Wet and dry bulb hygrometer

One of the concerns of meteorologists is to obtain a measurement of the amount of water vapour in the air. Two thermometers, one of which has a wet piece of muslin wrapped around its bulb, will give a value of the **relative humidity**. Relative humidity is a measure of how close to saturation the water vapour in the air is. The wet bulb will be **cooler than the dry one** owing to the cooling effect of the evaporation of the water. The drier the air the more evaporation will take place and therefore the cooler the wet bulb will be in relation to the dry one. Tables have been drawn up listing the relative humidity for dry bulb temperatures and the amount of cooling of the wet bulb.

Supersaturation

If the air is free of any small solid particles such as dust, water drops will not form, for the drop can only start to form when the vapour has condensed on such a particle. If there are no particles the air will become **supersaturated**. An example of this is found in **the cloud chamber** which is described on page 178 in the chapter on radioactivity.

Key terms

Thermometers measure the degree of hotness.
Celsius suggested the centigrade scale of temperature. Units are degrees Celsius (°C).
Absolute scale of temperature was first suggested by Kelvin.

Units are degrees Kelvin (K). The Kelvin temperature is found by adding 273 to the Celsius temperature.

Coefficient of linear expansion The fractional change in length for one degree rise in temperature. Units are /°C.

Unusual expansion of water Water contracts when heated from 0°C to 4°C.

Charles' law states that a gas at constant pressure will expand by $\frac{1}{273}$ of its volume for each rise in temperature of 1°C.

Pressure law states that the pressure of a fixed mass of gas at constant volume is proportional to its Kelvin temperature.

Conduction The transmission of heat, usually through solids, when the heat energy is passed on from one part of the substance to the next.

Convection The transmission of heat by a liquid or gas by the movement of the fluid.

Radiation The transmission of heat by electromagnetic waves.

Kilocalorie An old-fashioned unit of heat energy.

James Joule was one of the scientists who established that heat is a form of energy that can be changed to other forms. Thus it can be measured in joules (J).

Four-stroke internal combustion engine The petrol engine used in most motor cars, in which heat energy is changed into mechanical energy.

Heat capacity of a body is the energy required to raise the temperature of the body by 1°C.

Specific heat capacity The energy required to raise the temperature of 1 kg of a substance 1°C. Units are joules per kilogram per degree Celsius (J/kg°C).

Latent heat The energy that a substance needs to change from a solid to a liquid or from a liquid to a gas. It does not produce a change in temperature.

Specific latent heat The energy needed to change 1 kg of a substance from solid to liquid or from liquid to gas. Units are joules per kilogram (J/kg).

Refrigerators use latent heat (required for vaporisation) to move heat from one place to another. It takes heat from the inside of the refrigerator and transfers it to the air outside the cabinet.

Regelation The effect of decreasing the melting temperature of ice when the pressure on it increases.

Saturated vapour occurs when there is a state of dynamic equilibrium between a liquid and its vapour, i.e. the vapour cannot hold any more molecules escaping from the liquid.

Saturated vapour pressure The pressure produced by the saturated vapour. Its value increases considerably with increased temperature.

Relative humidity measures the amount of water vapour present in the air as a percentage of the total water vapour which would be required to produce saturation at the same temperature.
Hygrometer An instrument for measuring the relative humidity of the air.
Supersaturation The unstable state reached when air saturated with water vapour is cooled without condensation of any water. The air is then supersaturated.

Examples

Heat energy and specific heat capacity

A 200-watt immersion heater raises the temperature of 480 g of water from 18°C to 30°C in 2 minutes. Heat losses from the water are negligible. Find (a) how many joules of heat energy have been supplied; (b) the specific heat capacity of water.

On page 38, it was stated that power is the rate of doing work, or the rate of using energy, and that:

$$\text{power in watts} = \frac{\text{energy in joules}}{\text{time in seconds}}$$

Here we have a 200-watt heater operating for 2 minutes. Using the above relationship:

Energy in joules = 200 × 120 = 24 000 J.

When this energy is applied to the water, the temperature rises from 18°C to 30°C, i.e. by 12°C. There are 480 g of water. In S.I. one would normally work in kilograms. Since:

energy = mass × s.h.c. × rise in temperature, we have

$$24\,000 = 0{\cdot}480 \times \text{s.h.c.} \times 12$$

$$\therefore \quad \text{s.h.c.} = \frac{24\,000}{0{\cdot}48 \times 12} = 4\,167 \text{ J/kg°C.}$$

Notice carefully the units of s.h.c. In the calculation we divided energy in joules by mass in kg and the rise in temperature in °C. So the units of s.h.c. must be J/kg°C.

Suppose now that the same heater was immersed in 300 g of a liquid the specific heat capacity of which is 2 500 J/kg°C. What would be the rise in temperature in two minutes?

Here the basic energy equation has to be applied again. Since the same heater is used for the same time, 24 000 joules of energy are again supplied. Thus:

$$24\,000 = 0{\cdot}3 \times 2\,500 \times \theta, \quad \text{so } \theta = \frac{24\,000}{750} = 32°\text{C}$$

The method of mixtures

0·3 kg of liquid of specific heat capacity 2 500 J/kg°C at a temperature of 40°C is mixed with 0·2 kg of water of specific heat capacity 4 200 J/kg°C and temperature 20°C. What is the temperature of the mixture?

Let the temperature of the mixture be θ°C. The rise in temperature of the water is $(\theta - 20)$°C, and the fall in temperature of the liquid is $(40 - \theta)$°C.

Heat energy gained by water: $0{\cdot}2 \times 4\,200 \times (\theta - 20)$ joules

Heat energy lost by liquid: $0{\cdot}3 \times 2\,500 \times (40 - \theta)$ joules

Assuming that there is no loss of energy to the surroundings:

$$0{\cdot}2 \times 4\,200 \times (\theta - 20) = 0{\cdot}3 \times 2\,500 \times (40 - \theta)$$

$$\therefore \quad 840(\theta - 20) = 750(40 - \theta)$$

The arithmetic is simplified if we divide each side of the equation by the highest common factor of the terms outside the brackets before multiplying out. This h.c.f. is 30.

$$\therefore \quad 28(\theta - 20) = 25(40 - \theta)$$

$$\therefore \quad 28\theta - 560 = 1\,000 - 25\theta$$

$$\therefore \quad 53\theta = 1\,560 \quad \theta = \frac{1\,560}{53} = 29{\cdot}4°\text{C}$$

This is a fairly simple example of this type of calculation. In practice it is usually assumed that no heat is lost to the surroundings, but often it is worth taking into consideration the heat energy which would be transferred to the vessel, or calorimeter, in which the mixing was carried out. In the above example this would add an extra term to the left-hand side of the equation.

Latent heat

(1) In an experiment to measure the latent heat of fusion of ice, 6 g of dry ice at 0°C were added to 60 g of water contained in a calorimeter of water equivalent 8 g, and caused the temperature

of the water to fall from 18°C to 10°C. Taking the specific heat capacity of water to be 4 200 J/kg°C, calculate the specific latent heat of fusion of ice.

There are a number of points to note in this calculation. Firstly, the energy taken from the calorimeter cannot be ignored, but the correction made for this is a simple one.

The term **'water equivalent'** means that the calorimeter may be regarded as equivalent to an additional 8 g of water. So the calorimeter may be forgotten and the system regarded as being thermally equivalent to $60+8=68$ g of water. Therefore heat lost by water and calorimeter $= 0{\cdot}068 \times 4\,200 \times 8$ J. (There is no need to work this out at this stage.)

When the ice is placed in the water, it melts and then finally ends up as water at 10°C (the final temperature of the mixture). As far as the calculation is concerned, the gain in energy must be split into two separate parts:

(1) gain in energy of ice at 0°C in changing to water at 0°C = mass of ice × latent heat of fusion $= 0{\cdot}006 \times L$ joules.

(2) gain in energy of the melted ice changing in temperature from 0°C to 10°C = mass × specific heat capacity × rise in temperature $= 0{\cdot}006 \times 4\,200 \times 10$ joules.
Notice that 4 200 is used in the last expression, because once the ice has melted it becomes water, so the specific heat capacity of water is used.
Assuming that the heat lost by the calorimeter and contents is equal to that gained by the ice:

$$0{\cdot}068 \times 4\,200 \times 8 = 0{\cdot}006 \times L + 000{\cdot}6 \times 4\,200 \times 10$$

$$\therefore \quad 2\,285 = 0{\cdot}006\,L + 252$$

$$\therefore \quad L = \frac{2\,033}{0{\cdot}006} = 338\,833 \text{ J/kg.}$$

(2) Steam at 100°C is allowed to pass into water in a container until the temperature rises to 50°C. Both water and container are initially at 10°C. If the water and the container together have a heat capacity of 30 875 joules, calculate the mass of steam which changes into water. The latent heat of vaporisation of steam is 2 260 000 J/kg.

Energy lost by m kg of steam
(a) to change it into water at 100°C = $m \times 2\,260\,000$
(b) to cool this water to 50°C = $m \times 4\,200 \times 50$
Total energy lost by steam = $m(2\,260\,000 + 4\,200 \times 50)$
Energy gained by water and container = $30\,875 \times 40$
Energy lost = energy gained.

$$\therefore \quad m = \frac{30\,875 \times 40}{(2\,260\,000 + 4\,200 \times 50)}$$

$$= 0{\cdot}5 \text{ kg.}$$

$\therefore$ Mass of steam changed into water = 0·5 kg.

Calculation relating to specific heat capacity and specific latent heats

To calculate how much energy is required to change 2 kg of ice at −6°C into steam at 300°C, a five-part calculation must be made.

(1) Raise the temperature of the ice from −6°C to its melting point 0°C (specific heat capacity of ice is 2100 J/kg°C):

energy = 2 × 2100 × 6 J.

(2) Change the ice at 0°C to water at the same temperature (specific latent heat of fusion of ice is 336 000 J/kg):

energy = 2 × 336 000 J.

(3) Raise the temperature capacity of the water from 0°C to 100°C (specific heat of water is 4 200 J/kg°C):

energy = 2 × 4 200 × 100 J.

(4) Change the water at 100°C to steam at the same temperature (specific latent heat of vaporisation of water is 2 260 000 J/kg):

energy = 2 × 2 260 000 J.

(5) Raise the temperature of the steam from 100°C to 300°C (specific heat capacity of steam is 1 460 J/kg°C):

energy = 2 × 1 460 × 200 J.

To find the total energy required all five amounts of energy must be added together.

Chapter 4
Wave Motion

Water waves

Sound waves, light waves and radio waves are all important in everyday life. To understand them it is best to consider water waves first, to study their properties and then compare them with similar properties in other types of waves.

Frequency A vibrating dipper placed in water contained in a shallow container called a **ripple tank** can be made to produce **a number of waves every second**; this is the frequency, f, of the wave. The units of frequency are **hertz** (Hz).

Wavelength The distance between two waves (i.e. between two successive crests) is the wavelength, λ. The usual units of wavelength are **metres** (figure 41).

Velocity Each wave moves through the water with a fixed velocity, v. It does not vary with the height of the wave (the **amplitude**, a) nor with the frequency. The usual units of velocity are **m/s**.

The equation connecting frequency, wavelength and velocity is: **velocity = frequency × wavelength** or $v = f\lambda$

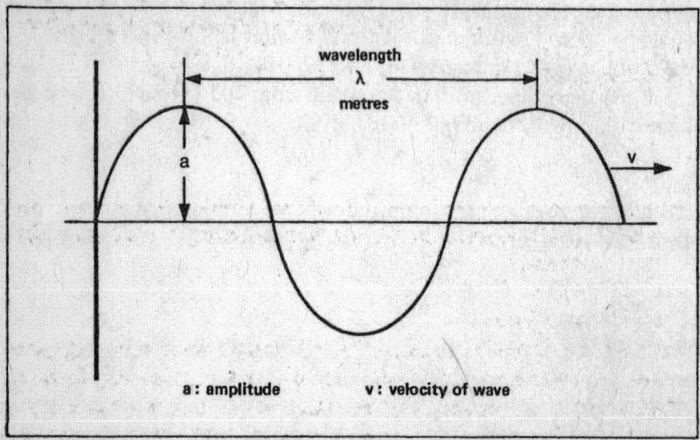

Figure 41. Wavelength and amplitude of a typical waveform

For the velocity to remain constant, when the wavelength decreases the frequency must increase by an equivalent amount. The two must be inversely proportional.

Reflection of water waves

If a barrier is placed in a water-filled ripple tank, waves hitting it will be reflected.

Note that in figures 42 and 43 the small arrows show that the direction of motion of the wave is always at right angles to the **wavefronts**, which may be considered as the crest of each wave.

In figure 42a a set of plane waves is hitting the barrier at an angle and being reflected from the barrier at a similar angle.

In figure 42b a set of circular waves is shown radiating from a source, *O*. After reflection the waves have the reverse curvature, and so appear to be coming from point *I*, which is as far behind the barrier as *O* is in front.

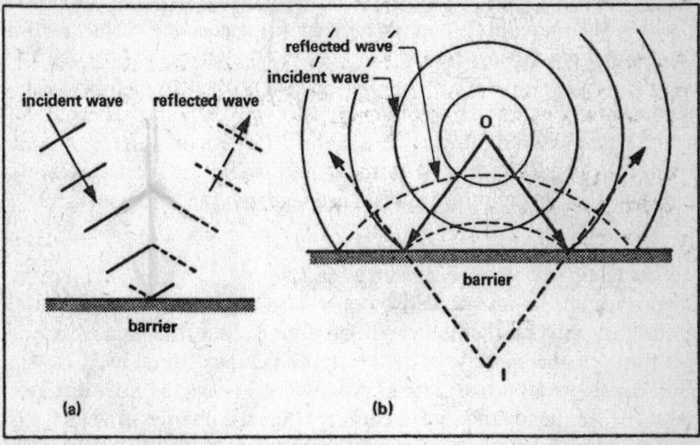

Figure 42. (a) Plane waves and (b) circular waves reflected from a plane barrier

Curved barriers

With the **concave** side of the barrier facing the oncoming plane waves a curvature is imposed upon the waves, making them all **converge** to a single point *F* (figure 43a). The reverse effect is produced when a circular wave started at this point produces a plane wave after reflection.

When a plane wave is reflected by a **convex** surface, the reflected wavefronts are curved, appearing to **diverge** from a point *F* behind the barrier (figure 43b).

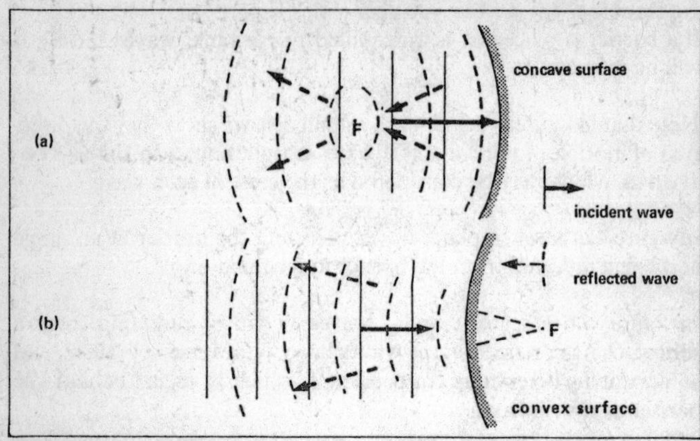

Figure 43. Reflection of plane waves by (a) a concave surface; (b) a convex surface

Among many examples of other types of waves being reflected off straight and curved surfaces are light reflected off mirrors, radar transmitting and receiving aerials, and radio-telescopes. Highly sensitive directional microphones use concave reflecting surfaces.

Refraction of water waves

If perspex sheets are placed in part of a ripple tank so that the water just covers them, it will be found that the water waves travelling in the shallow water over the perspex are slowed down. This **change in velocity** of a wave is known as **refraction**. Since the frequency of a wave cannot alter, the change in velocity is accompanied by a corresponding **change in wavelength**. This is shown in figure 44a. Note that here the approaching wavefronts are parallel to the boundary between the deep and shallow water. In figure 44b, the wavefronts **strike the boundary at an angle**, and the effect of refraction is to **change their direction of travel**. On emerging into deeper water, they regain their original wavelength, but since they struck the second boundary at an angle, they will again suffer a change of direction. In this case, the water waves are behaving in the same way as light when it passes through a prism.

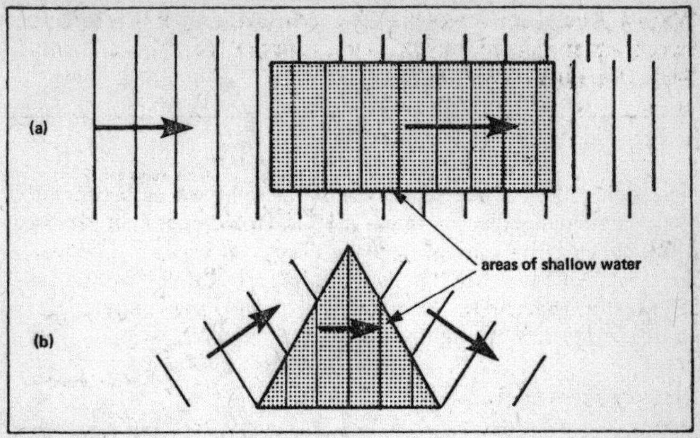

Figure 44. Refraction of water waves in shallow water: (a) normal incidence; (b) oblique incidence

Diffraction

This is the effect of **waves bending round corners** and **spreading out** after passing through narrow gaps (figure 45).

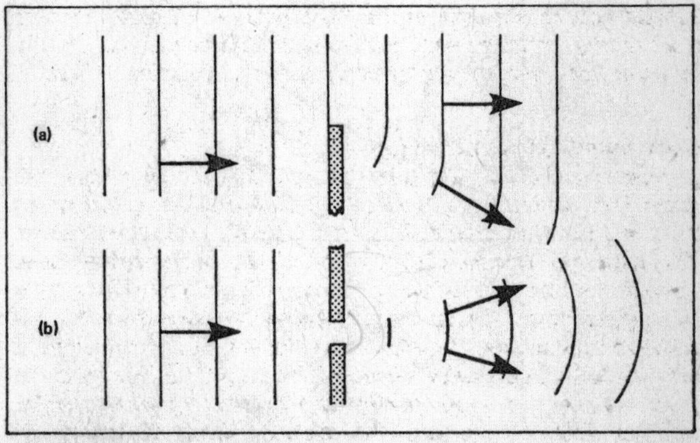

Figure 45. Diffraction of plane waves: (a) round the edge of a barrier; (b) through a narrow opening

To make a noticeable effect the wavelength has to be quite long. This is why sound waves easily bend round corners but light apparently does not. Similarly long radio waves, such as BBC Radio 2, with a wavelength of 1 500 m, have a good reception over a wide area, even in valleys sheltered from the transmitter. The receiving aerial for VHF, which has a much shorter wavelength, has to be in a direct line with the transmitter.

The same effect is not so noticeable in light waves because the wavelength of light is very much shorter. However, if light is passed through a narrow enough gap the effect can be seen. Look at a bright street-lamp through a gap made by your two thumbs. Gradually decrease the size of the gap and just before the gap closes completely, the light will seem to spread out.

Interference

Two sets of waves of the same frequency are produced using two dippers attached to the same motor. These sets of waves spread out from their sources, points *A* and *B* (figure 46) and soon overlap or **interfere** with each other.

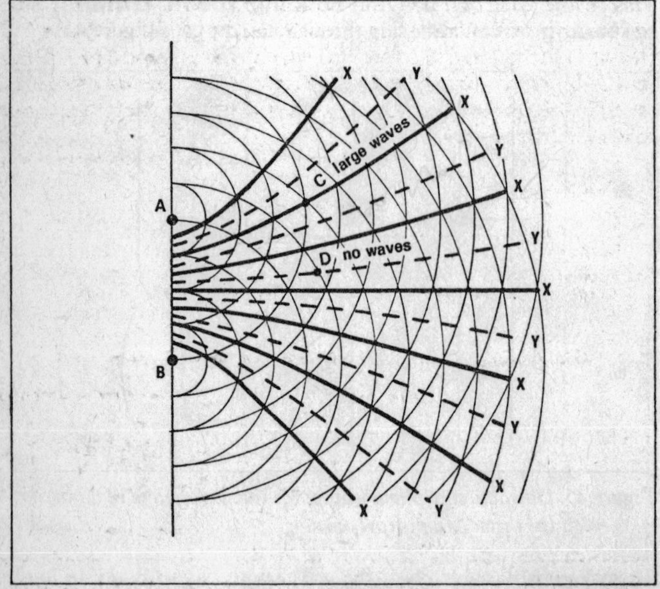

Figure 46. Interference of two sets of circular waves

Should two wavecrests arrive at the same place at the same time, as at *C*, they will combine to produce a larger wave. This is **constructive interference**, and is true at all points along the solid lines, where the two sets of waves are said to be **in phase**.

However, should a wavecrest from one source and a trough from the other arrive at the same place at the same time, as at *D*, they will combine to produce nothing. This is **destructive interference**, and is true at all points along the broken lines, where the two sets of waves are said to be **out of phase**.

Remembering that the whole pattern is continuously moving to the right of the diagram, waves arriving at the points *X* will have a large amplitude, double that of the original waves produced at *A* and *B*, while no waves at all will arrive at the points *Y*. Had these been light waves, the points *X* would appear bright on a screen, whereas the points *Y* would be dark.

Interference of sound waves

An interference effect similar to that just described can be demonstrated with the apparatus shown in figure 47. As the microphone is moved along the line *AB* the amplitude of the sound wave detected by the microphone and displayed on the screen of the oscilloscope varies from almost zero to twice that obtained if one of the speakers is blocked out. The experiment is best done in the centre of the laboratory and as high above the benches as clamp-stands will allow, to minimise unwanted echoes. Best results are obtained at frequencies of several kHz.

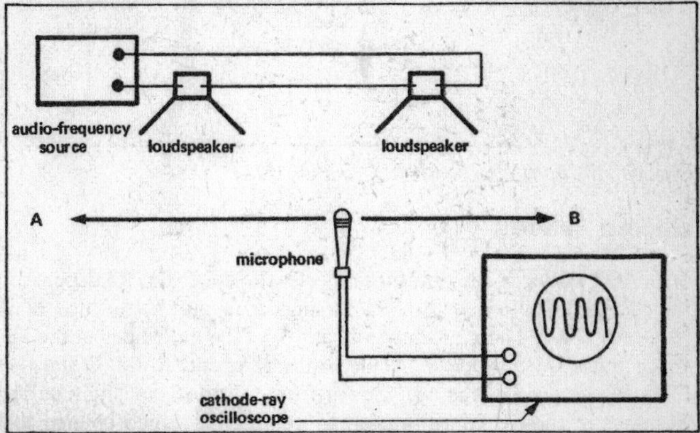

Figure 47. Detection of interference of sound waves from two identical sources

Interference of light waves: Young's slits

To bring about interference in light waves the first task is to produce two **coherent sources**, that is, sources emitting light of the same frequency and in phase with each other. Young first did this by passing a narrow beam of light through two slits very close together.

The two narrow beams that are thus produced must now be made to overlap. If light travelled in straight lines they would never do so. However, diffraction spreads them out slightly and by putting a screen about 2 metres away **alternate bright and dark bands** can be seen, corresponding to positions where the two sets of waves arrive at the screen in phase and out of phase, respectively (figure 48).

The spacing of the successive bright bands, or **fringes**, increases with the wavelength of the light used. Thus red light would give more widely spaced fringes than blue light, since red light has a longer wavelength.

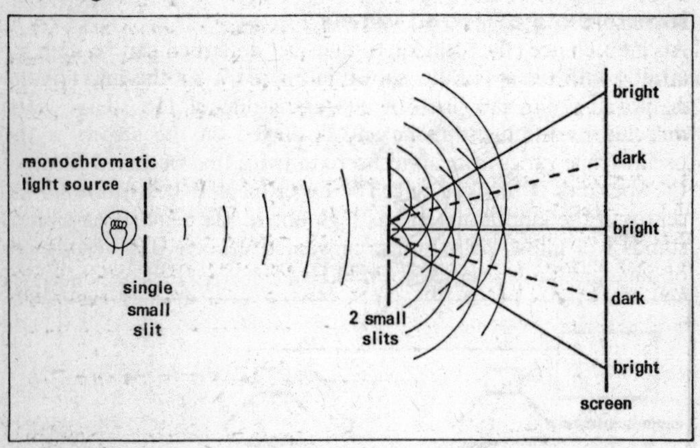

Figure 48. Interference of light: Young's slits

Sound waves

Sound is produced by vibrating objects: the cone of a loudspeaker, the prongs of a tuning fork, the strings in a guitar, the lips of a trumpet player. These produce vibrations of the molecules of the air that is passed from one molecule to the next and so on. When the disturbance reaches the ear, the eardrum vibrates, causing a series of bones in the ear to vibrate on to nerve ends which change the mechanical vibrations into electrical vibrations which are then transmitted to the brain.

Longitudinal and transverse waves (figure 49)
Sound waves are **longitudinal**, each molecule vibrating to and fro parallel to the direction in which the wave is travelling. Light waves on the other hand are **transverse**; the electromagnetic vibrations are at right angles to the direction of motion of the wave. A slinky spring can be made to vibrate in either of these ways by pushing the end in and out, or shaking it to and fro. Water waves can be considered as transverse. A floating cork moves up and down in the same place while the wave advances.

Velocity of sound
From everyday experience we know that sound travels more slowly than light. Thunder is heard after the lightning is seen, a cricket ball is seen to be hit before it is heard, the flame from a starting pistol is seen before the report is heard.

If we try to measure the velocity of sound by timing the delay between seeing the flash and hearing the report of a starting pistol, two difficulties arise. The first is that we are assuming that the light takes no time to travel (a fair assumption, in fact). The second is the problem of measuring such as short time, which is comparable with the reaction time of the experimenter.

To overcome both these difficulties an echo method can be used that produces a result of fair accuracy. The experimenter stands a measured distance, say 50 metres, from a large building, and makes a series of sharp claps; he will hear the echoes coming back off the wall of the building. With practice he will be able to make the

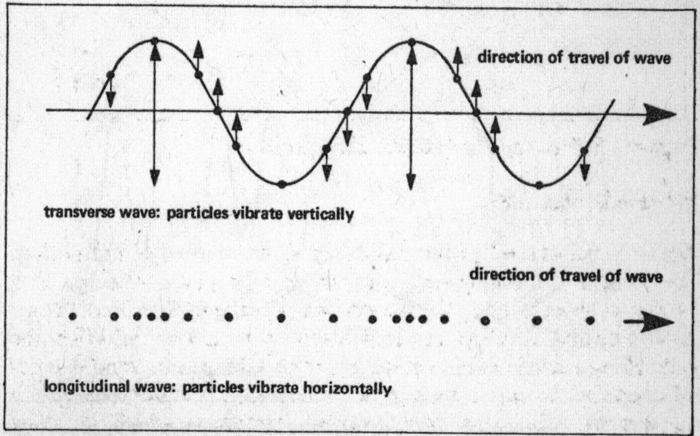

Figure 49. Transverse and longitudinal waves

clapping coincide with the returning echo. When the experimenter has achieved this rhythm his partner can time, say, twenty claps, and calculate the time for one clap to travel to the wall and back again.
The value obtained is about 330 m/s.

The media through which sound travels

Sound requires a substance through which to travel. For example, if an electric bell is enclosed in a bell jar, it can be shown that as the air is removed from the jar the intensity of the sound will decrease until almost no sound at all can be heard.

Sound will also travel through a liquid: echo-sounding equipment demonstrates this. Sound travels much faster in water (about 1 500 m/s) than it does through air.

Sound will also travel through solids. If a watch is placed on one end of a bench, it will be possible to hear it ticking by listening at the other end of the bench several metres away.

Musical sounds

A generally held view is that music is sound which is pleasant to listen to, as opposed to noise, which is unpleasant. Therefore music to one man's ears may be noise to another's! Usually, however, music consists of a series of notes each with a fixed frequency, whereas noise is a jumble of frequencies (figure 50).

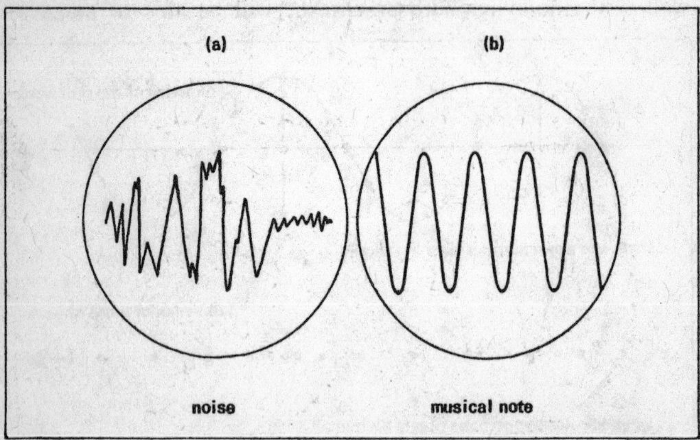

Figure 50. Typical waveforms of (a) noise; (b) a musical note, displayed by a cathode-ray oscilloscope

Pitch and frequency

A musical note has a **higher pitch**, and is higher in the musical scale, if it has a **higher frequency of vibration**. Middle C on the pianoforte has a frequency of 262 Hz: the note one tone higher, D, has a frequency of 294 Hz. In changing from one note to another an octave above the frequency is doubled. Thus the note C - an octave above middle C, has a frequency of 524 Hz.

The lowest note of the pianoforte has a frequency of 27·5 Hz and the highest 3 520 Hz. Young people can hear sounds up to a frequency of about 20 kHz, but as people get older this upper frequency drops considerably. Certain animals have a higher limit: for example, dogs can hear a whistle that humans cannot. Bats use their high-pitched whistle as a kind of radar. The returning echo tells them of the presence of objects in front of them.

Loudness

A musical note of fixed frequency can be played loudly or softly. It is the **amplitude** that is being changed. A loud note has a large amplitude, a quiet note a small amplitude (figure 51a).

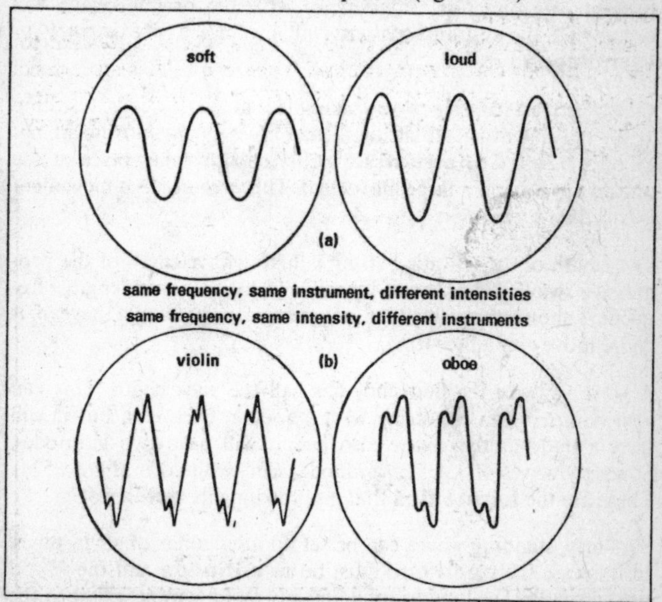

Figure 51. Typical waveforms: (a) a soft and loud note; (b) sounds of different qualities

Quality of sound

Notes played by two different musical instruments can have the same frequency and the same loudness, yet it is not difficult to hear the difference between, say, a violin and an oboe. Though they are playing notes with the same basic frequency, the shape of the wave each emits is different. This can be seen from the trace that is produced on the C.R.O. screen (figure 51b). The difference is caused by the **overtones**, or **harmonics**, that are emitted along with the basic frequency. Because instruments have different shapes and different methods of producing sound, each instrument will produce harmonics of different intensities. The human ear is quite capable, with a little training, of picking out these differences.

Standing waves

If one end of a piece of string is shaken to and fro, a wave pattern will be seen to travel or progress along the length of the string. This is an example of a **progressive** wave. If however the string is stretched and fixed at both ends, the progressive wave reaching one of the ends will be reflected and travel back up the string in the opposite direction. Interference with the original wave will occur, and the resulting pattern will produce a **stationary** or **standing wave**.

Destructive interference takes place at the two ends, resulting in no movement of the string. These areas of no movement are called **nodes**. **Constructive interference** takes place at the centre, resulting in a large movement. This area of large movement is called an **antinode**.

The length of the vibrating string is half a wavelength of the progressive wave. The same formula (velocity = wavelength × frequency) shows the velocity a progressive wave would have if it could move along the string.

A wave of twice the frequency has half the wavelength. This can also be fitted into the string with a node at each end, but it will have a node at the centre also and it will have two antinodes. Similarly waves of 3, 4, 5 ... antinodes can be fitted in (figure 52a). These are the **harmonics** that give the tone its **quality**.

Similarly, standing waves can be set up in columns of air in pipes. In this case the **open end** must be an **antinode** and the **fixed end** a **node**. The fundamental fits only $\frac{1}{4}$ of a wavelength into the pipe and the first harmonic fits $\frac{3}{4}$ of a wavelength into it. Thus the first harmonic is 3 × fundamental frequency (figure 52b). It should be

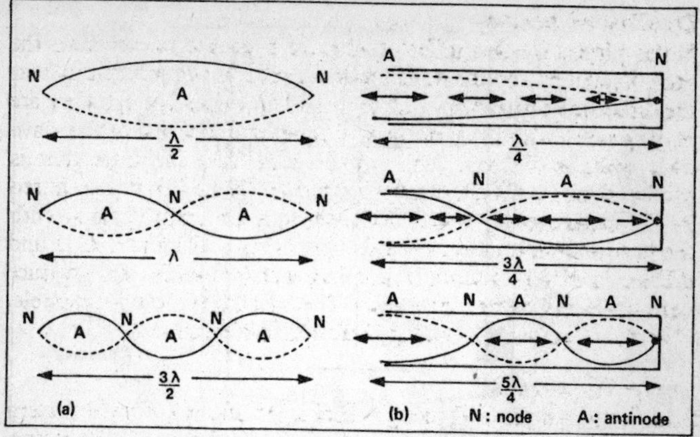

Figure 52. 1st, 2nd and 3rd harmonics produced by standing waves (a) in a stretched string; (b) in a pipe closed at one end

remembered that the air molecules are vibrating parallel to the length of the pipe.

Resonance

Most people have blown over the top of a bottle or test-tube and heard the whistle of fixed frequency that is produced. The pitch of the note can be increased by putting water into the tube. The tube has its own natural frequency that depends on its length. If a note of the same frequency is sounded close to the tube using a tuning fork or an audio-oscillator and loudspeaker, the air in the tube is set in vibration. We say that the tube **resonates**.

Resonance will also occur if the driving source is at the frequency of one of its harmonics. At other frequencies the tube will not respond.

Experiment for measuring the velocity of sound in air using a resonance tube

A tuning fork of known frequency is needed. The tuning fork is vibrated over the mouth of a tube whose effective length can be varied by raising or lowering the level of water in the tube (figure 53). Starting with the tube full of water the length is slowly increased till resonance is heard. The length of the tube is measured, and further increased until it is about three times this first length. A second resonance length is found.

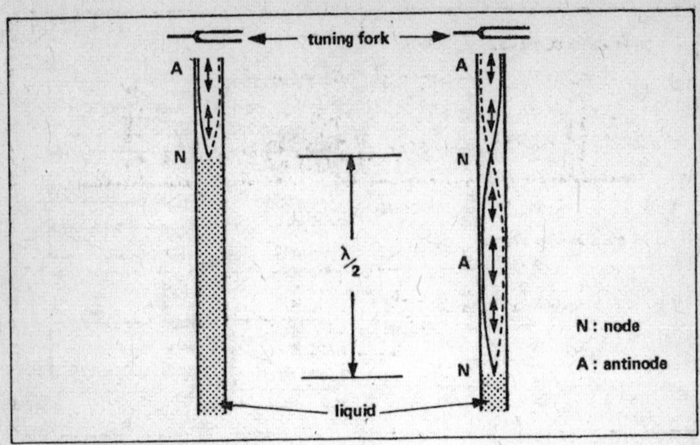

Figure 53. Resonance in a tube showing 1st and 2nd position of resonance

Standing waves are set up at resonance. The first length corresponds to a quarter wavelength, $\lambda/4$. The second length corresponds to three-quarters of a wavelength, $3\lambda/4$. To each length must be added a small unknown end correction, since the antinode is not exactly level with the open end of the tube. However, the difference between the two will be half a wavelength, $\lambda/2$.

Since the frequency of the fork is known, the velocity of the wave can be calculated using the formula $v = f\lambda$.

Other examples of resonance

Pushing a child's swing illustrates mechanical resonance. A series of small pushes at the correct frequency will produce a swing of large amplitude.

Another example is shown in the figure, **a trolley held in position with springs**. The system has a natural frequency and small pushes at this frequency will produce large oscillations (figure 54a). The value of the natural frequency can be decreased by increasing the mass of the trolley or by reducing the stiffness of the springs.

The tuning circuit of a radio illustrates electrical resonance. Radio waves with many frequencies are present in the air and picked up by the aerial. The tuned circuit will resonate to only one frequency depending on the values of the inductance and capacitance

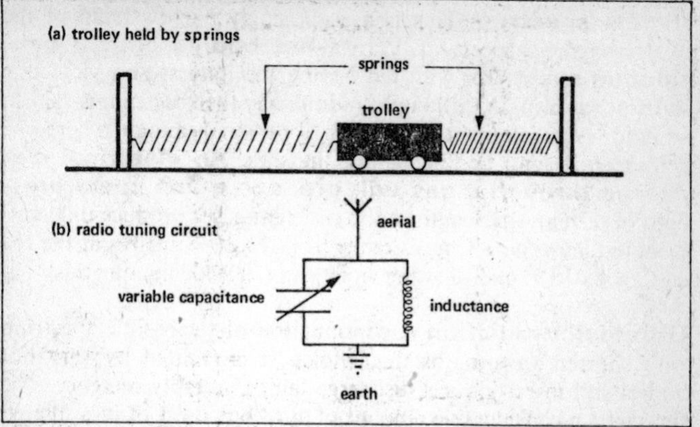

Figure 54. Systems illustrating resonance: (a) mechanical system; (b) electrical system

(figure 54b). When a radio is tuned the value of the capacitor is changed and when the waveband is changed another inductance is being switched in.

The family of electromagnetic waves

These are **transverse progressive waves** consisting of electric and magnetic vibrations at right angles to each other and at right angles to the direction of travel of the wave. They do not need a substance through which to travel for they can travel through a vacuum. They all have the same speed in vacuum, of $\mathbf{3 \times 10^8}$ **m/s**.

Radio waves These are mostly man-made, with a range of wavelengths from thousands of metres down to a few centimetres. Each radio station has its own frequency and wavelengths (Remember that frequency × wavelength = velocity, that is 3×10^8 m/s.) The sources of radio waves in the universe are studied by **radio-astronomers** using **radio-telescopes**.

Infra-red radiation This is emitted by hot bodies and is similar to visible light but has a longer wavelength. When absorbed by a surface the energy which it carries is changed to heat energy. The presence of heat energy is one way of detecting infra-red radiation. Wavelengths range from about 1 millimetre down to the point at which they merge into the red end of the visible spectrum at about 7×10^{-7} metres or 700 nanometres (nm).

Visible spectrum This is a surprisingly narrow band of the electromagnetic spectrum. **White-hot bodies emit a continuous spectrum** with red having the longest and violet the shortest wavelength. Although traditionally the spectrum is said to contain seven colours it is a matter of personal opinion where one colour starts and another colour finishes. **An electrical discharge through a gas will produce a line spectrum**—light of certain wavelengths only. Different gases produce different-coloured lines. Wavelengths range from 7×10^{-7} metres at the red end, to 4×10^{-7} metres at the violet end (700–400 nanometres).

Ultraviolet radiation A continuation of the visible spectrum with shorter wavelengths than violet. It is emitted by very hot bodies and in certain gas discharge lamps, notably mercury. The sun emits a considerable amount of 'u.v.' but most of it is filtered out by the atmosphere before it reaches ground-level. It is u.v. that causes sunburn, the effect of which is much more noticeable in high mountain regions than at sea-level. Wavelengths range from about 4×10^{-7} metres to 10^{-7} metres (400–100 nanometres).

X-rays and gamma rays These have the highest frequency, or shortest wavelength, of all the electromagnetic waves. X-rays are emitted from metals when bombarded with high-energy electrons. Gamma rays are emitted from radioactive sources, and are very penetrating. Wavelengths range from 10^{-7} to 10^{-14} metres.

Key terms

Frequency is the number of vibrations per second. Units are hertz (Hz).
Wavelength is the distance between identical parts of adjacent waves. Units are metres (m).
Velocity of a wave is the distance any part of the wave-front moves in unit time. Units are metres/second (m/s).
Amplitude is the maximum displacement of each particle from its central position.
Reflection of waves takes place such that the angle of incidence is equal to the angle of reflection.
Refraction takes place when the wave velocity changes. Usually the direction of the wave is also changed.
Diffraction is the effect of waves bending round corners and spreading out after passing through narrow gaps.
Interference takes place when two waves of the same frequency pass through each other. **Constructive** interference occurs if the

waves are **in phase** and **destructive** interference occurs if they are **out of phase**.

Transverse waves The vibrations causing these are at right angles to the direction the wave is travelling.

Longitudinal waves The vibrations causing these are parallel to the direction of travel of the wave.

Pitch is a musical term which describes the frequency of a note. A higher-pitched note has a higher frequency.

Loudness of a noise or musical note is determined by the amplitude of the wave. The louder the noise the greater the distance through which each particle of air is vibrating.

Quality of a musical note tells the listener what instrument is playing the note. It is associated with the shape of the wave and depends on the combination of harmonics which are played with the fundamental note.

Standing waves are the result of the interference of two waves travelling in opposite directions in stretched strings and air pipes.

Nodes are places of no movement in a standing wave.

Antinodes are places of large movement in standing waves.

Resonance occurs if a vibrating system is set into motion by a second vibration of the same frequency.

Electromagnetic waves are transverse waves of electric and magnetic vibrations. They all have the same velocity, 3×10^8 m/s, in a vacuum. In order of increasing frequency they can be classified as follows:

(1) **Radio waves** Mainly man-made. Though of very small intensity, they are produced naturally in the heavens.

(2) **Infra-red waves** These have frequencies slightly lower than that of visible light, and are mainly emitted by hot bodies.

(3) **Visible spectrum** Red is the lowest frequency through orange, yellow, green and blue to violet, as the frequency increases.

(4) **Ultraviolet waves** These have frequencies slightly higher than that of visible light, and are emitted by very hot bodies and some gas discharge lamps.

(5) **X-rays** These are emitted by metals when they are bombarded by high-speed electrons.

(6) **γ-rays** These are waves emitted by radioactive elements.

Example

In figure 55, *A* is a source which sends out sound waves of frequency 110 Hz in all directions. At *B*, two sets of waves arrive, those which have travelled directly along *AB*, and those which have been reflected at *C* by a wall. Taking the speed of sound to be 330 m/s, calculate the wavelength of the waves.

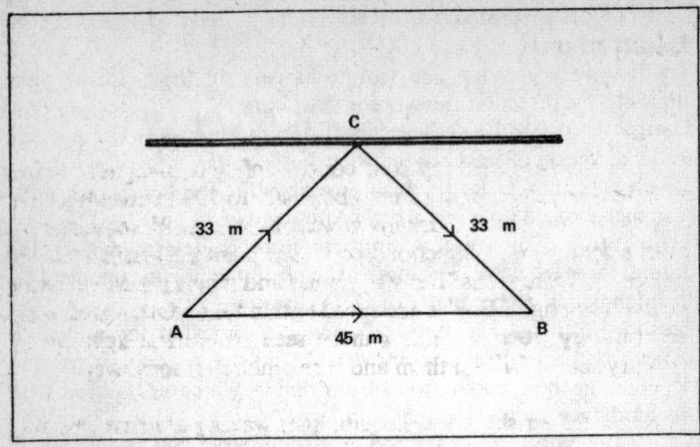

Figure 55. Interference of sound resulting from reflection

What is the time interval between the arrival at *B* of the direct wave and the reflected wave? What is the path difference between the two wave paths from *A* to *B*? What would happen if *A* were moved a short distance towards *B*?

The first part is straightforward: $\lambda = v/f = 330/110 = 3$ m.

To calculate the time taken for the sound to travel from *A* to *B* requires application of the formula: speed = distance/time. The extra distance which the reflected wave has to travel is $66-45 = 21$ m, so the extra time required is $21/330 = 7/110$ seconds.

The path difference is normally expressed in wavelengths. The difference in distance between the two paths is 21 m, and since the wavelength is 3 m, it follows that 21 m corresponds to exactly 7 wavelengths.

Since the path difference is an exact number of wavelengths, it follows that the direct wave and the reflected wave will arrive in phase at *B*. Thus the intensity of the sound at *B* will be a maximum.

If the source is moved towards *B*, the two sets of waves will no longer arrive in phase so the intensity of the sound will be reduced.

Chapter 5
Light

Light is a form of energy that consists of electromagnetic waves with wavelengths ranging from about 400 to 700 nanometres. The direction of the waves is usually indicated by rays. Waves travel out from a source in all directions, and rays point outwards from the source in all directions. The sun, flames and lamps are all **primary sources** of light: light is produced within them. Other bodies are **secondary sources** and can be seen only when light from a primary source falls on them and is re-emitted in some way.

Because of their short wavelength, light waves penetrate very little round corners and, except for work where a high degree of accuracy is required, it can be considered that **light travels from a source in straight lines through any continuous medium.**

Shadows and eclipses

The shadow produced by an opaque body when light from a point source falls on it will have clearly defined edges (figure 56a). However, light from an extended source will form a blurred shadow, **gradually** changing from a complete shadow to no shadow at all. For

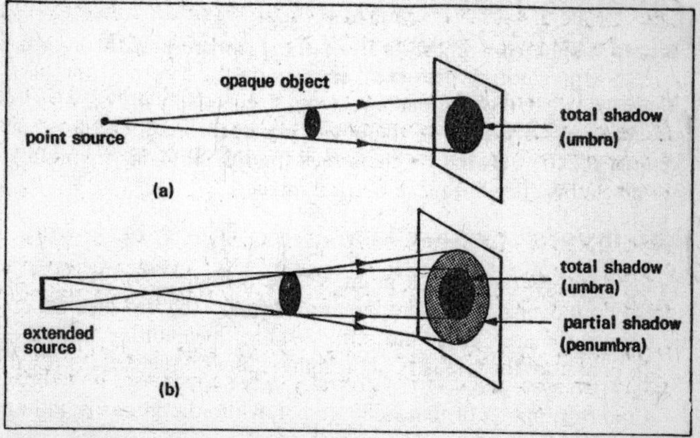

Figure 56. Shadow formed with an opaque object by light (a) from a point source; (b) from an extended source

the extended source the rays from the two extremes of the source to the extremes of the opaque object are drawn; other rays fit in between (figure 56b). The region of total shadow is called the **umbra**, the region of partial shadow is the **penumbra**.

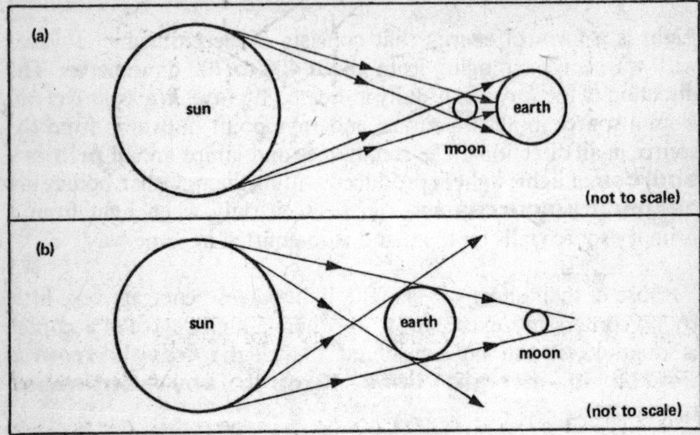

Figure 57. (a) Solar eclipse; (b) lunar eclipse

In an eclipse of the sun, or moon, the sun is the extended source, and the sun, moon and earth are in line. For a **solar eclipse** the shadow of the moon falls on the earth (figure 57a). For a **lunar eclipse** the shadow of the earth falls on the moon (figure 57b).

The pinhole camera

As rays of light from objects in front of the camera pass through the hole, they produce an **inverted image** on the screen at the back of the box. Since the pinhole is very small, effectively only a **single ray** from **each point** on the object passes through it to form an **image of that point** on the screen (figure 58). A larger pinhole will give a brighter but more blurred image.

The inverse square law

As light travels out from a point, source S (figure 59), the energy spreads out over a larger and larger area (twice the distance means four times the area, three times the distance means nine times the area). Therefore the intensity of the light falling on the surface B will be a quarter, or $\frac{1}{2^2}$, of that falling on A, while the intensity falling on C will be one-ninth, or $\frac{1}{3^2}$, of that falling on A. Hence the

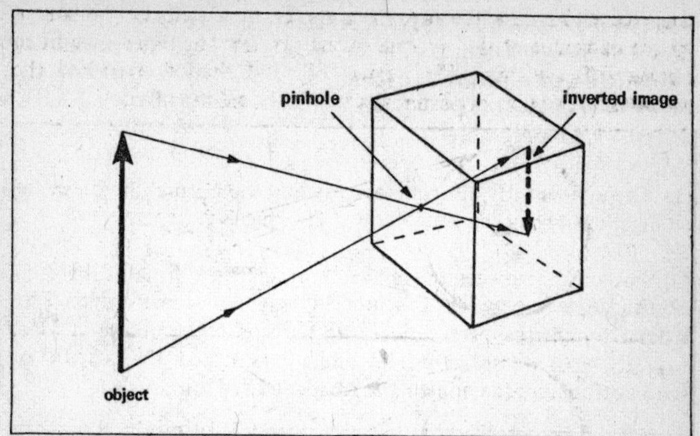

Figure 58. Formation of an image by a pinhole camera

intensity decreases by a factor corresponding to the **inverse of the square of the distance** or $\frac{1}{d^2}$, where d is the distance from the source.

This applies not only to light but to other phenomena which spread out in three dimensions, such as gravitational and magnetic forces.

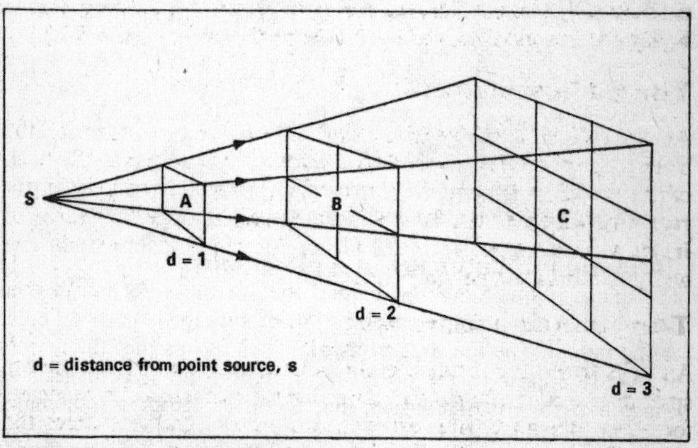

Figure 59. Illustration of inverse square law

Reflection of light

Since light is a form of electromagnetic radiation, it is reflected by highly polished surfaces, as discussed in the previous chapter. A

mirror is an example of such a reflecting surface. The laws which apply to the reflection of light are best explained by considering **reflection at a plane surface**.

Laws of reflection

(1) The incident ray, the reflected ray and the normal are all in the same plane. (They can all be drawn on a piece of paper.)

(2) The angle *r* between the reflected ray and the normal is the same as the angle *i* between the incident ray and the normal. **The normal** is the line perpendicular to the surface where the incident ray strikes it (figure 60). The angle *i* is called the **angle of incidence**, and the angle *r* the **angle of reflection**.

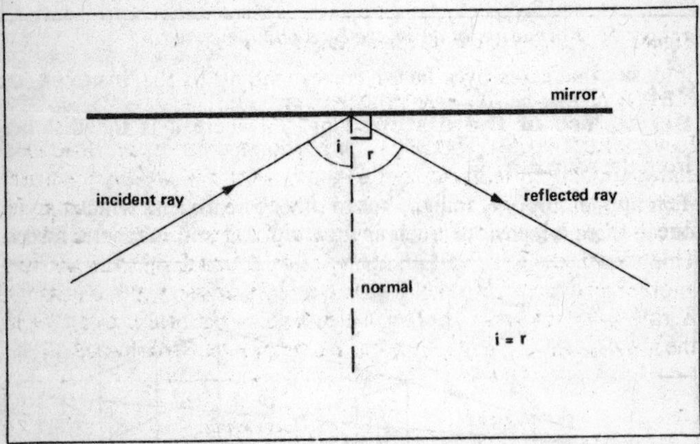

Figure 60. Reflection at a plane surface

Formation of an image in a plane mirror

The image is situated as far behind the mirror as the object is in front of it. A line joining an object and its image is perpendicular to the mirror. The image is **virtual**, which means that the rays of light only appear to the observer to come from the image. The image is **laterally inverted**, that is, the image of a right hand looks like a left hand.

Figure 61 shows an eye looking at the image of a point source. The eye imagines that the rays come from *I*. Note that the imaginary rays behind the mirror are dotted. It can be proved by geometry that $OA = AI$ and that OI is perpendicular to the mirror.

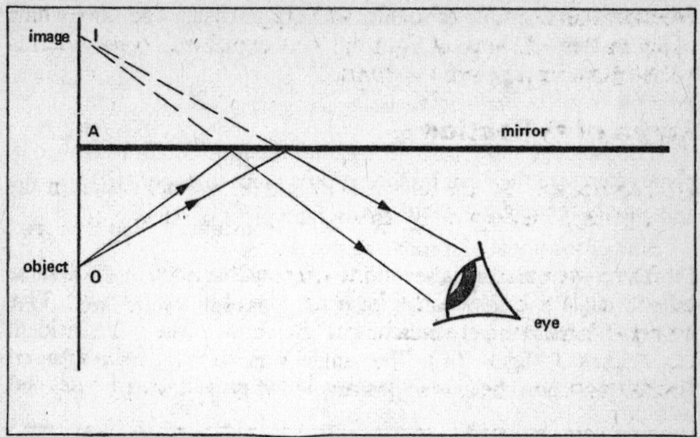

Figure 61. Image formation by a plane mirror

Refraction at plane surfaces

It was stated earlier that light travels in straight lines through a continuous medium. At the boundary between two media, however, there is a change in velocity which usually results in a change in direction, unless the ray is travelling at right angles to the boundary. This effect is called refraction. A ray of light travelling from air into another medium, e.g. into glass or water, is bent towards the normal. A ray travelling from another medium into air bends away from the normal. Angle *i* in figure 62 is the **angle of incidence**, angle

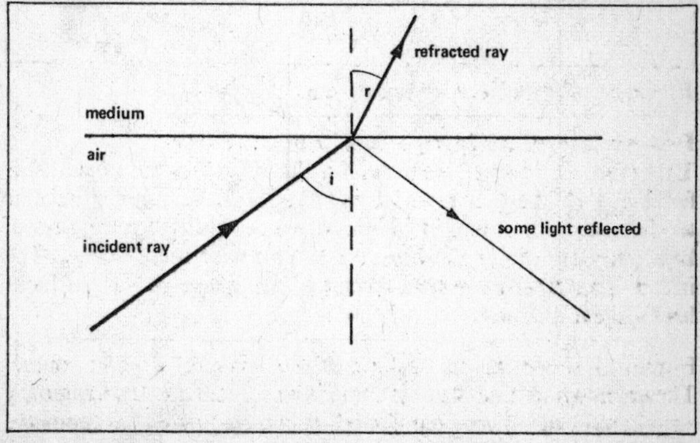

Figure 62. Refraction

r the **angle of refraction.** Note that there is also a certain amount of light reflected.

Laws of refraction

(1) The incident and refracted rays and the normal are in the same plane, therefore they can be shown on a piece of paper.

(2) $\frac{\text{Sine of the angle of incidence, } i}{\text{Sine of the angle of refraction, } r}$ = a constant, called the **refractive index**, n. This second law, named after its discoverer, is called Snell's law. The values of n for glass and water are about 1·5 and 1·33 respectively.

It can be shown that the refractive index, n, can also be defined as follows:

$$n = \frac{\text{the velocity of light in air}}{\text{the velocity of light in the medium}}$$

Apparent depth

The depth of a swimming-pool looks less than it really is. A glass block on a piece of paper makes the paper appear higher. These and other such effects can explained in terms of refraction.

A ray of light from an object O bends away from the normal at the surface and into the observer's eye (figure 63). The observer, not realising that the light has bent, thinks it comes from I.

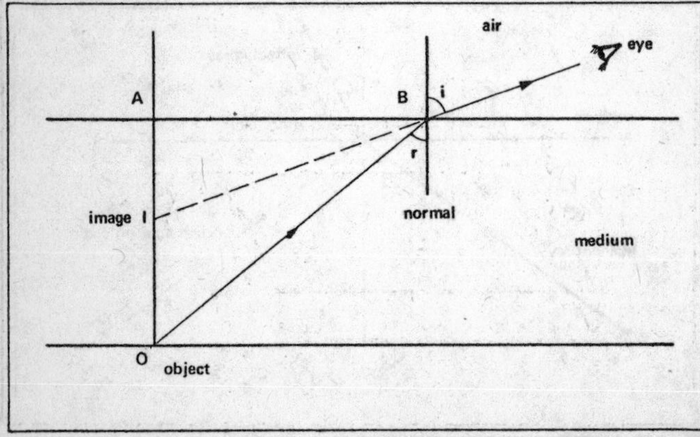

Figure 63. Apparent depth

I is therefore a virtual image of O. Since the ray would follow the same path if its direction were reversed, we can consider angle $AOB = r$, and $AIB = i$ therefore

$$OB = \frac{AB}{\sin r} \quad \text{and} \quad IB = \frac{AB}{\sin i}$$

Dividing the first equation by the second:

$$\frac{OB}{IB} = \frac{\sin i}{\sin r} = n.$$

If the eye is moved so that it is almost vertically above O

then $OB = OA$ and $IB = IA$ thus $n = \dfrac{\text{real depth}}{\text{apparent depth}}$

This gives a convenient method of finding n for a glass block or for a liquid.

To find the refractive index of a glass using real and apparent depth

A pin O is placed touching the far side of a glass block (figure 64). A second pin I, is moved backwards or forwards along the block till it appears to coincide with the image of O viewed through the block. This can be achieved accurately by a method of **no parallax. Parallax** is the **apparent relative movement** of two stationary objects at different distances from the observer when the observer moves. If there is no parallax, no apparent movement, the two objects must be the same distance from the observer. Thus I coin-

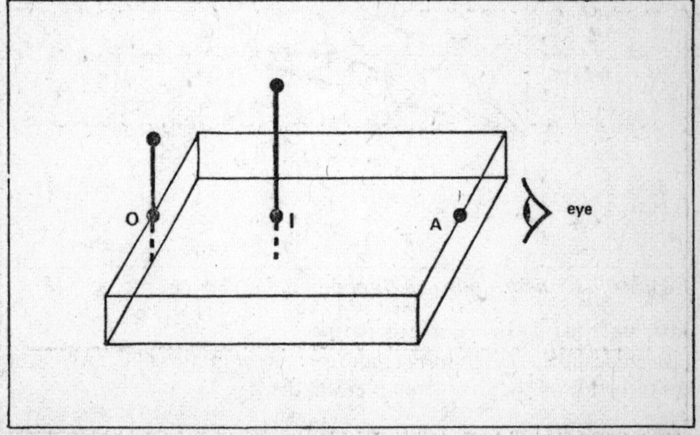

Figure 64. The refractive index of glass using real and apparent depth

cides with the apparent position of *O* when there is no parallax between them as the eye moves from side to side. The distances *AO* and *AI* can be measured and *n* calculated since *AO* represents the real depth and *AI* the apparent depth.

Total internal reflection

Total internal reflection occurs when light travelling from another medium into air strikes the boundary at an angle **greater than the critical angle**, *c*.

Consider the light rays from *O* (figure 65). *OW* strikes the surface normally and most of it passes straight through. A small proportion is reflected. *OX* strikes the surface at an angle of incidence less than *c* and most of the light passes through to the air, bending away from the normal. A proportion is reflected.

OY is called the critical ray; it makes an angle *c* with the normal, called the **critical angle**. Some of the light passes into the air and skims along the surface, so in the air the ray makes an angle of 90° with the normal. The rest of the light is reflected. *OZ* makes an angle with the normal greater than *c*, none is refracted, the whole is reflected, total internal reflection occurs.

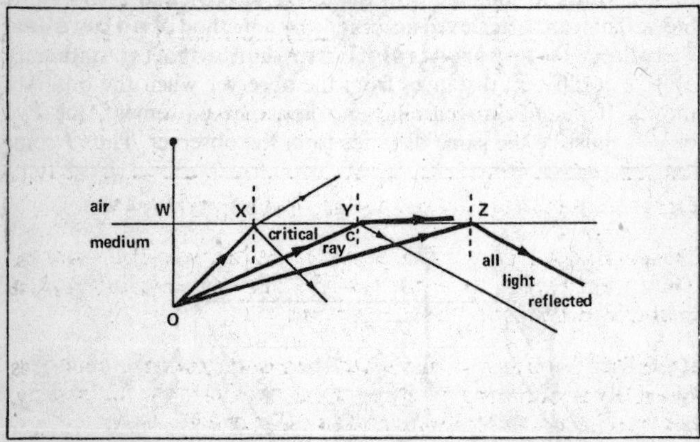

Figure 65. Critical angle and total internal reflection

The value of the critical angle

The angle in the medium is *c*, and the angle in air is 90°. Again, if the direction of the ray were reversed, then:

$$\frac{\sin i}{\sin r} = \frac{\sin 90^\circ}{\sin c} = n, \quad \text{but } \sin 90^\circ = 1$$

$$\text{Therefore } \frac{1}{\sin c} = n \quad \text{or } \sin c = \frac{1}{n}$$

For glass, $n = 1{\cdot}5 \quad c = 42°$

For water, $n = 1{\cdot}33 \quad c = 48°$

Total reflection prisms

These prisms have two angles of 45° and the other of 90°. When the light passes through one of the shorter faces normally, it strikes the longer face at an angle of 45° to the normal. This is more than the critical angle for glass, therefore the light is totally reflected, and passes out through the third face normally (figure 66a).

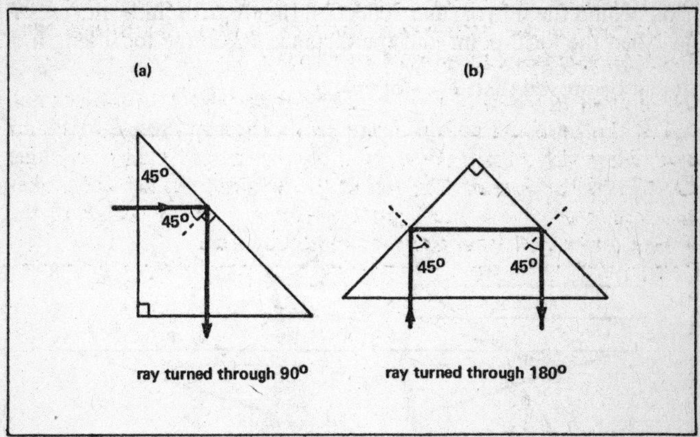

Figure 66. Total internal reflection in 45° right-angled prisms

Two such prisms are used in periscopes rather than plane mirrors. They reflect a greater proportion of the light and there are no ghost images as can occur with plane mirrors.

If the light is incident on the longer face of the prism, it undergoes two internal reflections, finally emerging parallel to the incident ray, but travelling in the opposite direction (figure 66b).

Two pairs of prisms are used in this way in prismatic binoculars. The first pair turns the rays through 180° in the horizontal plane. The next pair also turns the rays through 180° but this time in the vertical plane. This causes the image to be inverted, but since the binocular lenses by themselves give an inverted image the two inversions cause the final image to be the correct way up.

Reflection at curved surfaces

The concave and convex mirrors shown in figure 67 can be considered as parts of spheres with centres of curvature *C*, and radii *r*. The centre of the mirror is the pole, *P*. The line through *CP* is the principal axis.

For a concave mirror rays parallel and close to the principal axis will pass through a single point, *F*, on the axis after reflection (figure 67a).

For a convex mirror rays parallel and close to the principal axis will diverge and appear to have come from a single point, *F*, on the axis behind the mirror after reflection (figure 67b). In both cases *F* is called the focal point and the distance *PF* is the focal length, *f*.

It can be proved that: $f = \frac{r}{2}$ or $r = 2f$.

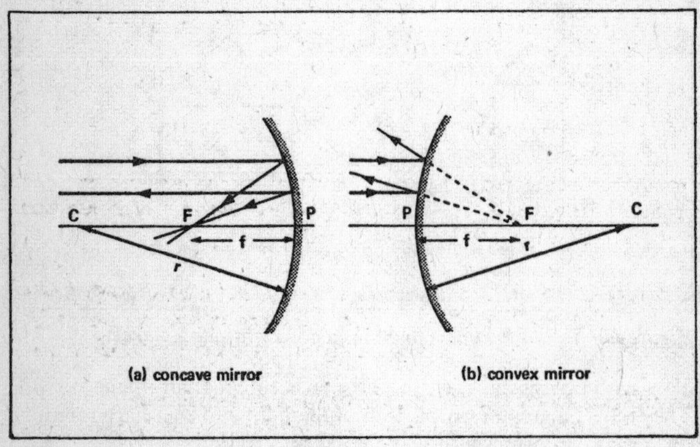

Figure 67. Reflection of parallel rays by (a) concave mirror; (b) convex mirror

Image formation by curved mirrors

Any line through *C* from the mirror will be the normal to the mirror at that point. Each part of the mirror obeys the laws of reflection, therefore the incident and reflected rays make equal angles with the line through *C* (figure 68). A **real image** is formed at the point where all the rays from an object **converge** after reflection. A **virtual image** is formed at the point from which all the rays from an object **appear to diverge** after reflection.

Thus in figure 68a, a real image of the object at O is formed at I, while in figure 68b a virtual image of O is formed at I. In general, the paths of two rays are sufficient to locate the position of the image; in this case the two rays are those reflected at P and Q. It can be shown that an object O will produce an image I such that: $\frac{1}{u}+\frac{1}{v}=\frac{1}{f}$ where u and v are the distances of the object and image from P.

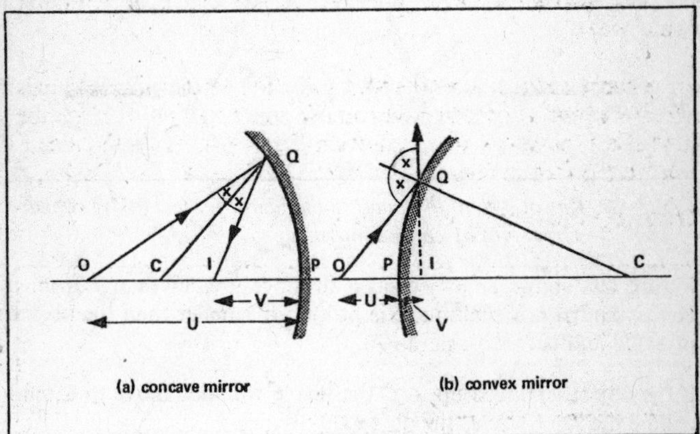

Figure 68. Image of a point on the principal axis (a) of a concave mirror; (b) of a convex mirror

For a convex mirror the numerical value of f is taken as **negative**, a virtual distance behind the mirror. Similarly if v works out to be negative this shows that a virtual image is formed, behind the mirror.

Graphical method of locating image position

Two rays of light are drawn from the top of the object. Where these two rays cross after reflection gives the position of the top of the image.

The two rays usually chosen are as follows. For a concave mirror a ray parallel to the principal axis passes through the focus after reflection. A ray passing through the centre of curvature strikes the mirror normally and returns along its own path (figure 69a).

For the convex mirror the ray parallel to the principal axis appears to come from the focus after reflection. A ray directed towards the centre of curvature is reflected back along its own path (figure 69b).

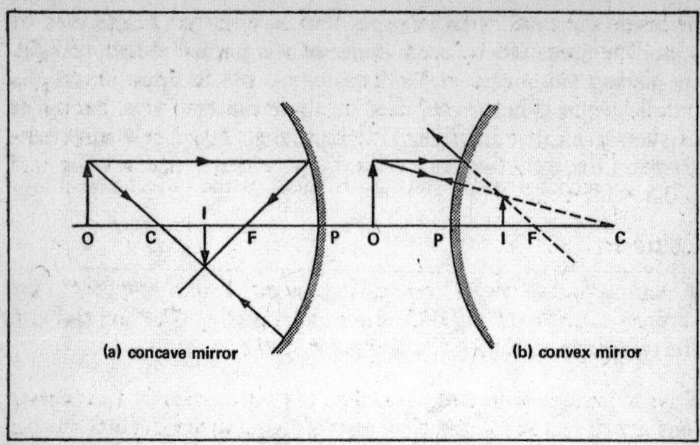

Figure 69. Ray diagrams for locating the image formed (a) by concave mirror; (b) by convex mirror

Figure 69a shows an object at a distance greater than *r* from a concave mirror. A real image is produced, smaller than the object, inverted and between *C* and *F*.

If the object is moved up to *C* the image will also move to *C*, and will be the same size as the object.

As the object moves from *C* to *F* the image moves further away from *C*, and will get bigger and bigger. When the object is at *F* the rays will not meet, but will be parallel.

If the object is between *F* and *P* the image will be virtual, behind the mirror, upright and enlarged.

Figure 69b shows an object in front of a convex mirror. Wherever the object is placed, a smaller, upright, virtual image is formed behind the mirror between *P* and *F*.

Uses of curved mirrors

Convex mirrors These include the mirrors used in shops to deter shoplifters, as well as wing mirrors on motor vehicles. The person using it has a much wider field of view but sees a smaller image than would be shown by a plane mirror.

Concave mirrors These are used for shaving and make-up mirrors, for which the object to be magnified (the user's face) must be inside the focal point in order that an enlarged image may be seen. They can also be used to produce a parallel beam of light, by placing the source at the focal point. In the opposite way, a parallel beam of light is reflected to a point at the focus. Examples of such concave mirrors include searchlights and reflecting telescopes. Concave reflectors are also used in sound and in radio and radar equipment.

Lenses

A lens is made up of transparent material with two spherical surfaces of centres C_1 and C_2 and radii r_1 and r_2. The line through the two centres is called the **principal axis**.

Rays of light close to and parallel to the principal axis of a convex lens are refracted so that they pass through a single point on the axis (figure 70a). Rays of light close and parallel to the axis of a concave lens appear to diverge from a single point (figure 70b). These points are called the **focal points**. Each lens will have a focal point on either side of the lens.

The point bisecting the two focal points is the **optical centre** of the lens; this may or may not be the physical centre. The distance between the optical centre and a focal point is called the **focal length** of the lens. There is no simple relationship between the radii of curvature and the focal length, for it also depends on the refractive index of the material.

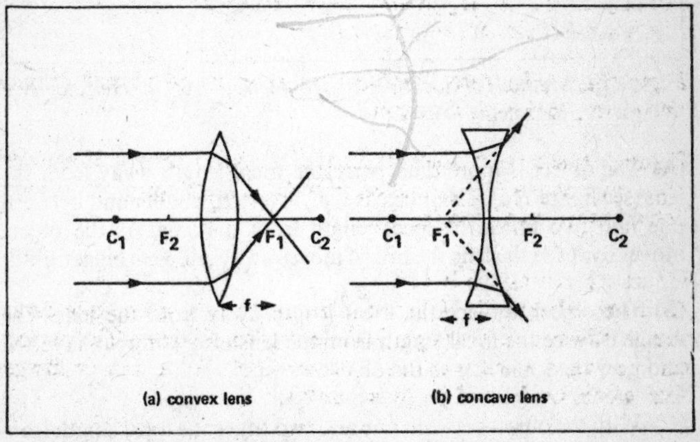

Figure 70. Position of the focal points of (a) convex lens; (b) concave lens

Graphical method of finding the position of an image

The paths of two rays from the top of the object are known. Where the two rays intersect after passing through the lens gives the position of the top of the image.

(1) The ray through the optical centre goes straight through undeviated.

(2) The ray parallel to the principal axis passes through the focus after passing through the lens in the case of a convex lens (figure 71a) and appears to have come from the focus in the case of a concave lens (figure 71b).

For a convex lens there are **five cases** to consider:

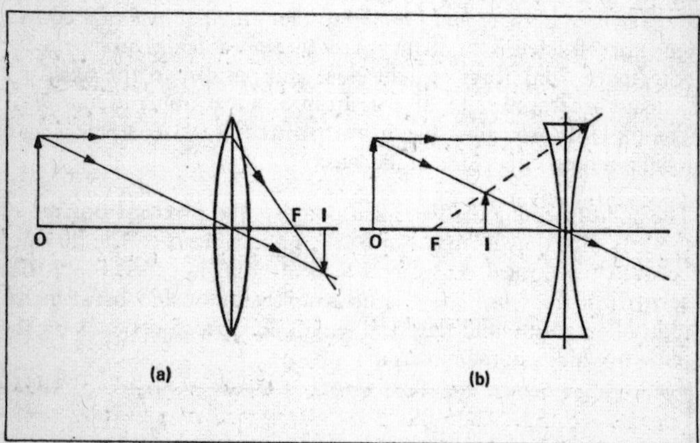

Figure 71. Method of locating image formed by (a) convex lens; (b) concave lens

(A) The object is more than twice the focal length away from the lens (see figure 71a). The image is real, inverted, smaller and between one and two times the focal length from the lens. As the object moves towards the lens the image moves away and gets bigger until:

(B) The object is twice the focal length away from the lens. The image is twice the focal length from the lens also, still real, inverted and now the same size as the object.

(C) With the object between one and two times the focal length from the lens the image is more than twice the focal length from the lens, still real and inverted, and now larger than the object.

(D) When the object is at the focal length, after refraction the rays are parallel. The image can be considered as being formed at ±infinity, real or virtual, inverted or upright, and very large.

(E) With the object inside the focal length the image becomes virtual, the rays diverge and have to be traced back. It is on the same side of the lens as the object, upright and larger.

For a concave lens the object can be any distance from the lens, and a virtual, upright, diminished image is formed on the same side of the lens as the object.

Practical details The lens is considered as a single vertical line for the drawing of the rays.

Choose a suitable scale and note it down. The larger the diagram the more accurate the results. The vertical scale need not be the same as the horizontal scale. If the size of the object is not given, make it some convenient height.

Formulae for lenses

If the distance between the object and lens is u, the distance between the image and the lens is v and the focal length of the lens f, then:

$$\frac{1}{u}+\frac{1}{v}=\frac{1}{f} \qquad \text{Magnification} = \frac{\text{size of image}}{\text{size of object}} = \frac{v}{u}$$

For concave lenses the focal length is taken as negative. Values of v that work out as negative indicate virtual images on the same side of the lens as the object.

Experimental methods of finding the focal length of a convex lens

(1) A simple, quick method, though not very accurate, is to produce the image of a distant object on the screen. Since the rays from the distant object are almost parallel the image is at the focal point, therefore the distance between the screen and the lens provides the focal length.

(2) A plane mirror behind the lens produces an image alongside the illuminated object. As the rays are retracing their paths they must be striking the mirror normally and thus they must be parallel. The arrangement is as case D above: the object is at the focal point, the distance between the object and the lens provides the focal length (figure 72a).

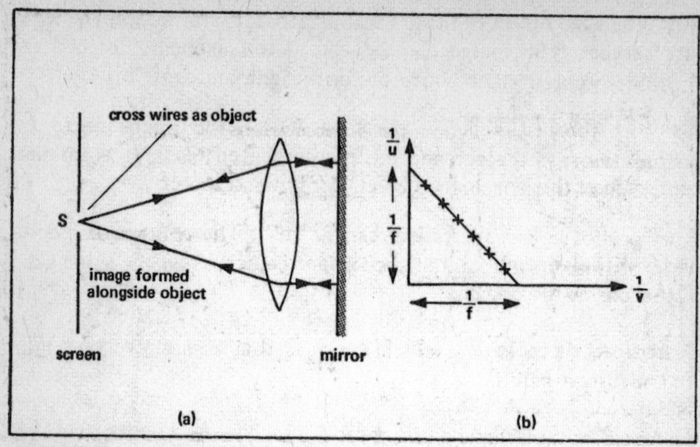

Figure 72. Methods of measuring the focal length of a convex lens

(3) For a more accurate method values of u and v are obtained using an illuminated object and a screen. A graph is drawn of $1/u$ against $1/v$ which gives a straight line having equal intercepts on each axis equal to $1/f$ (figure 72b).

Practical application of lenses: the camera and the eye

The **camera** uses a convex lens to produce a real, inverted, smaller image on the light-sensitive film, case A, page 110. A clear image is obtained for objects at different distances from the camera by moving the lens closer to, or farther from, the film, usually by a screw device which is marked for various distances. The **shutter**, normally closed, is opened briefly to allow light to fall on the film. The **aperture** allows the width of the beam of light entering the camera to be varied. This is controlled by an **iris diaphragm**. The interior of the camera is painted black to absorb any stray light, and the casing must be light-tight (figure 73a).

The size of the aperture and the speed of the shutter have to be considered together. For a moving object the shutter-speed must be fast but a large aperture will be necessary to allow sufficient light on to the film. On the other hand, to obtain a clear picture of objects at various distances from the camera at the same time (a large **depth of field**), the aperture should be as small as possible. This will mean that a slow shutter-speed is required for correct exposure of the film.

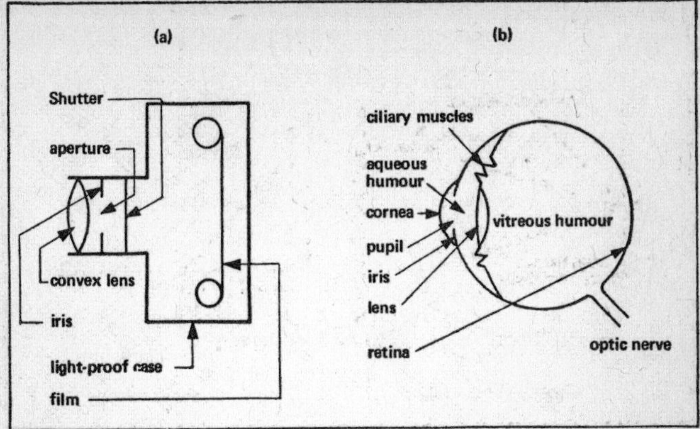

Figure 73. (a) the camera; (b) the eye

The **eye** functions in a similar way to the camera. A similar image is produced on the **retina**, the light-sensitive screen on the back of the eye. The image is conveyed to the brain by the **optic nerve**. The image is inverted but the brain corrects this.

Light is refracted through the front curved surface, the **cornea**, and is further refracted by each surface of the lens. The **iris** adjusts the size of the **pupil** to allow the correct amount of light into the eye. Outdoors on a bright day the pupil is very small, but in poor light it opens wide.

Accommodation, which is the ability of the eye to see objects at different distances, is achieved by the **ciliary muscles** which control the shape of the lens. For viewing close objects the focal length of the lens must be reduced, so the lens is made fatter. For viewing distant objects, the lens is made thinner. Accommodation and the adjustment of the iris are both automatic.

Defects of vision

With normal eyesight a person can accommodate objects from a great distance away to a near point of 25 cm or so.

Short sight or **myopia** The eye can see clearly objects closer than 25 cm but distant objects are blurred. For these distant objects the lens cannot be made thin enough and the image is formed in

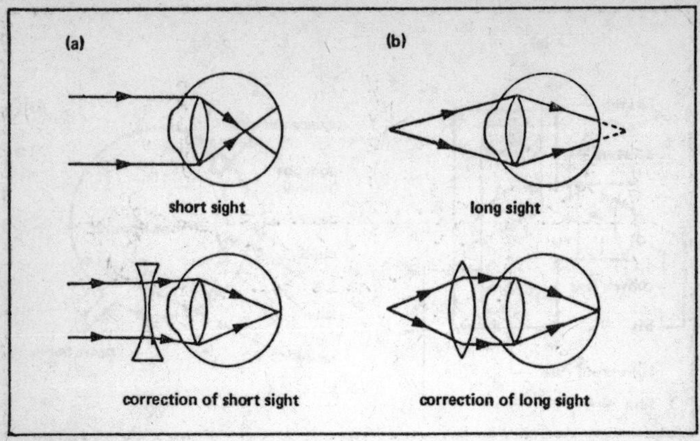

Figure 74. Correction of defects of vision: (a) short sight; (b) long sight

front of the retina. This is corrected by placing a concave lens in front of the eye. Figure 74a shows such an eye looking at a distant object. Remember that rays of light from a distant object are almost parallel.

Long sight or **hypermetropia** The eye can see distant objects but its near point is considerably greater than 25 cm. For objects closer than this the lens cannot be made fat enough and the image is formed behind the retina. This is corrected by placing a convex lens in front of the eye (figure 74b).

Simple microscope or magnifying glass

A short-focal-length convex lens is used with the object placed just inside the focal point, case E, page 111. The image is produced at the near point.

Compound microscope

This consists of two lenses, **both of short focal length**. The object is placed just outside the focal point of the **objective lens**. This forms an enlarged, real image, case C, page 110. This intermediate image is the object for the **eyepiece lens** and falls just inside the focal length of this lens. This again is as case E above and the image is again produced at the near point. Thus both lenses produce a certain magnification (figure 75).

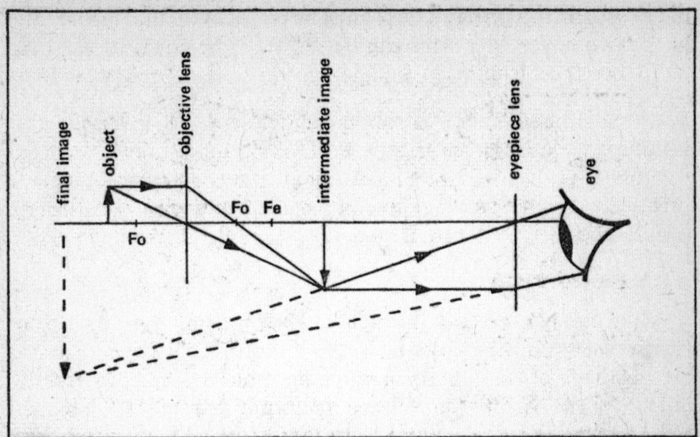

Figure 75. The compound microscope

Astronomical telescope

This also consists of two lenses, but here **one is of long focal length and the other short. The objective lens** is of long focal length and it produces a small inverted image of the distant object at its focus. This intermediate image is the object for the short-focus **eyepiece**. It is formed at the focal point of the eyepiece lens. Thus we have case D, page 111. The final image is also a great distance from the lens.

The previous definition of magnification cannot be applied here and

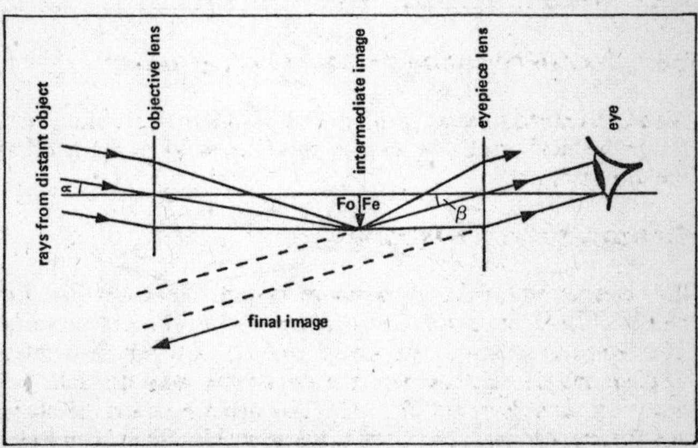

Figure 76. Astronomical telescope

the **angular magnification** must be considered, that is, the ratio of the two angles, β/α. This can be shown to be equal to the ratio of the two focal lengths, *fo/fe* (figure 76).

The final image of the astronomical telescope is inverted. This produces little inconvenience when looking at stars but is a considerable disadvantage when looking at things on earth. A pair of prisms can be used as suggested on page 105 or a short-focal-length lens can be used as in case B, page 110, to re-invert the image.

The spectrum

If white light is passed through a glass prism, it splits up, or **disperses**, into the colours of the spectrum. Dispersion occurs because each of the colours in white light has a different refractive index. In the visible part of the spectrum red has the smallest refractive index, and is bent the least. Violet has the highest refractive index and is bent the most. Orange, yellow, green, blue and indigo are between these two extremes.

Production of a pure spectrum

In the simple case described above, although the colours are, to a certain extent, split up, there is still an overlapping of neighbouring colours. In order to produce a pure spectrum, the arrangement shown in figure 77 is used. The first lens produces a parallel beam of white light from a point source at its focus. The prism splits it up so that the red comes out parallel but at a different angle from the other colours. The second lens collects each colour into a single spot on its focal plane.

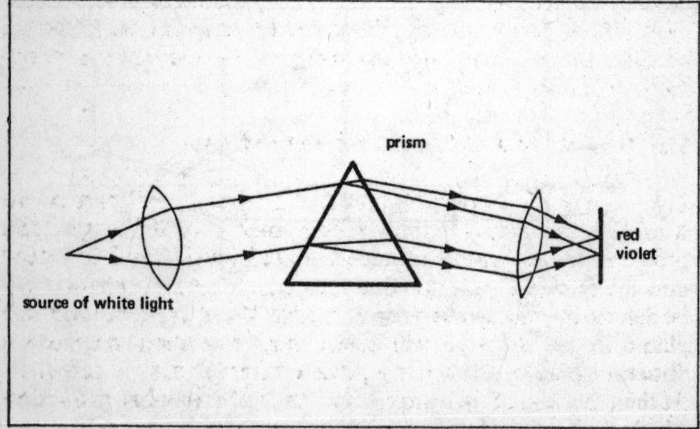

Figure 77. Production of a pure spectrum

Colour

Colour addition occurs when two or more beams of coloured light are shone together on a white screen. In theory, if the whole spectrum of colours is shone on to the screen it appears white. In practice it is found that only three colours are necessary, **blue**, **green** and **red**, the **primary colours**. It is also found that mixture of these three beams in varying intensities will produce, fairly well, any other colour. This is the theory of colour television: three electron beams strike patches of three substances which emit these three colours.

Colour subtraction is concerned with pigments and dyes. A red article looks red in white light because it reflects only red light and absorbs all other colours which may fall on it. It subtracts, takes away, all the colours but red. So in theory, if the red article is viewed under any colour but red it will absorb that colour and appear black. In practice the absorption is not complete and colours on either side of the spectrum are also emitted slightly. This produces the theory of pigment-mixing for painting and colour printing. Only three colours need be used, **cyan**, **yellow** and **magenta**. These are the **secondary colours** and are produced when combinations of two of the primary colours are projected on to a white screen:

red + green = yellow;
red + blue = magenta;
blue + green = cyan.

Thus yellow absorbs all colours except red and green, and cyan absorbs all colours except blue and green. A mixture of yellow and cyan will therefore reflect only green. In a similar way, various combinations of yellow, magenta and cyan can be mixed to give any required colour.

Infra-red and ultraviolet radiation

Light is a form of wave motion and is a very small part of the whole electromagnetic spectrum (see chapter 4). If the spectrum produced by sunlight is examined carefully, it is found to extend into the infra-red and ultraviolet regions. Infra-red radiation can be detected by its heating property: a blackened thermometer bulb placed in the infra-red will give a higher reading. Similarly a **thermocouple** attached to a galvanometer will show a deflection. At the other end of the visible spectrum is **ultraviolet radiation** which can be detected by a photographic emulsion such as that used for normal black and white prints.

Key terms

Primary sources Bodies which produce and emit light energy.
Secondary sources Bodies which reflect or scatter light energy.
Solar eclipse occurs when the moon moves between the sun and the earth, so that the shadow of the moon falls on the earth.
Lunar eclipse occurs when the earth moves between the sun and the moon, so that the shadow of the earth falls on the moon.
Inverse square law describes the manner in which light, gravitational force, magnetic force etc., spread out in three dimensions from the point of their source. Twice the distance yields one quarter the intensity, three times the distance yields one ninth the intensity.
Normal The line perpendicular to the surface where the incident ray strikes it.
Real image The image produced if the rays of light pass through the image position.
Virtual image The image produced if the rays of light appear to come from the image position.
Lateral inversion of an image means that the image appears the opposite way round but *not* upside down. Writing will appear back to front and a right hand will appear as a left hand.
Refraction The bending of light as it passes from one medium to another. It is caused by a change in the velocity of the light wave.

Refractive index $= \dfrac{\text{sine of the angle of incidence}}{\text{sine of the angle of refraction}}$

$= \dfrac{\text{velocity of light in first medium}}{\text{velocity of light in second medium}}$

Apparent depth An effect caused by refraction, such that:

$$\text{apparent depth} = \frac{\text{real depth}}{\text{refractive index}}$$

Total internal reflection occurs when all the light is reflected back into the medium and none is refracted.
Critical angle The angle of incidence just less than that at which total internal reflection takes place.
Radius of curvature of a curved mirror is the radius of the sphere of which the mirror is a part.
Principal axis is the line joining the centre of the mirror, the **pole** i.e. optical centre, and the **centre of curvature**.
Focal point of a concave mirror The point through which all rays parallel to the principal axis pass after reflection. It is half the radius of curvature of the mirror.
Focal point of a convex mirror The point from which all

rays parallel to the principal axis appear to diverge after reflection. Again, it is half the radius of curvature of the mirror.

Focal point of a convex converging lens is the point through which all the rays that are parallel to the principal axis pass after refraction by the lens.

Focal point of a concave diverging lens is the point from which all the rays parallel to the principal axis appear to diverge after refraction by the lens.

Retina The light-sensitive nerve network on the back of the eye.

Optic nerve carries the nerve impulses to the brain.

Cornea The transparent membrane over the front of the eye.

Iris The variable shutter that alters the size of the **pupil**.

Ciliary muscles control the shape of the lens allowing it to focus objects at various distances. This ability is called **accommodation**.

Myopia (short sight) Ability to focus only on objects which are a short distance away. Distant objects are blurred because the lens in the eye is too strong.

Hypermetropia (long sight) Ability to see distant objects but not to focus on close objects because the lens in the eye is too weak.

Simple microscope or **magnifying glass** A short-focal-length convex lens.

Compound microscope An instrument with two short-focal-length convex lenses, the **objective**, close to the object, and the **eyepiece**.

Astronomical telescope An instrument with two lenses: the **objective**, a weak convex lens of long focal length, and the **eyepiece**, a strong convex lens of short focal length.

Linear magnification $= \dfrac{\text{image size}}{\text{object size}}$

Angular magnification $= \dfrac{\text{angle object subtends at the eye}}{\text{angle image subtends at the eye}}$

Dispersion White light is split up into a spectrum of colours when it passes through a prism. This occurs because of the different wavelengths and refractive indices of the different colours.

Colour addition When two or more beams of coloured light are added, the eye does not see the two colours but a single colour, of a value between these two in the spectrum.

Colour subtraction The process whereby paints and dyes absorb certain parts of the spectrum of the light falling on them, taking away those colours and reflecting or scattering the remaining parts of the spectrum.

Example

Graphical solution of lens problem

An object is placed 80 cm from a screen. A converging lens is placed between them so that a clear image, magnified three times, is obtained on the screen. Find by scale drawing the distance of the lens from the object and the focal length of the lens.

The stages are as follows.

(i) Draw the principle axis.
(ii) Mark in a scale, and draw object and image 80 cm apart. Make the image three times the height of the object.
(iii) Join *AB*. *AB* represents the ray which passes through the optical centre of the lens, cutting the principal axis at *L*. The lens has thus been located.
(iv) Draw *AX* parallel to the principal axis.
(v) Join *XB*, *AX* represents a ray parallel to the principal axis, and *XB* is its path after refraction. *F* is therefore the principal focus of the lens.

By measurement, $OL = 20$ cm, and $LF = 15$ cm.

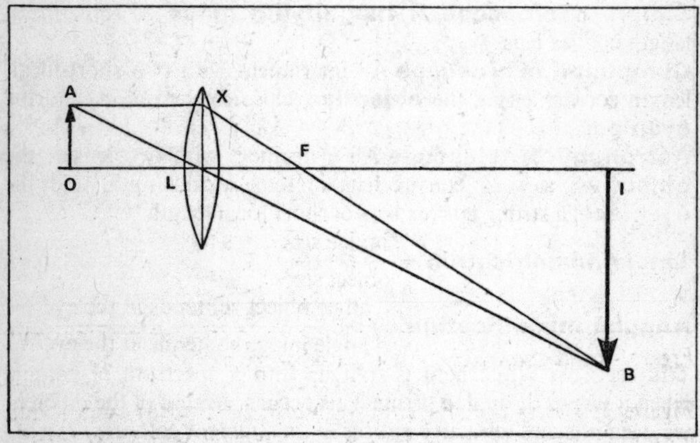

Figure 78. Graphical solution of lens problem

Chapter 6
Permanent Magnetism

Permanent bar magnets, when freely suspended, will point approximately north-south. The end pointing north is called the north pole, the other the south pole. If two such magnets are brought together with like poles next to each other they will repel each other. If unlike poles are brought together they will attract. **Like poles repel, unlike poles attract**. Substances that can be magnetised are iron, steel, nickel and cobalt and certain iron salts. Lodestone, or leading stone, which was known in ancient times to have these properties, is an iron oxide, magnetite (Fe_3O_4). The area around a magnet where its effects are felt by other magnetic materials is called its **magnetic field**. The shape of a magnetic field can be plotted using a small compass.

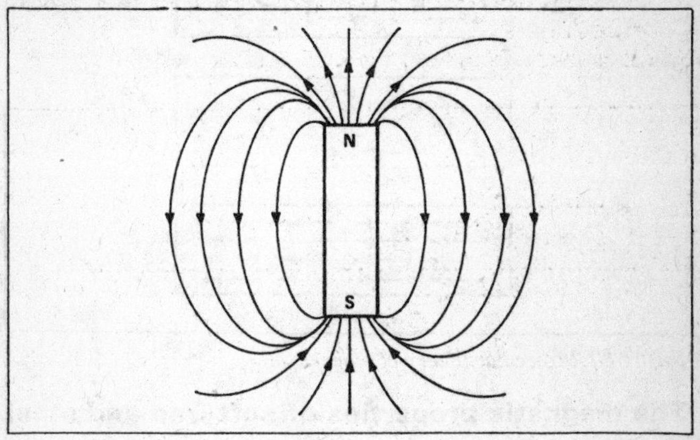

Figure 79. Magnetic field created by a bar magnet

Figure 79 shows the shape of the magnetic field around a bar magnet. The arrows on the lines, which are called **magnetic flux lines**, show the direction in which a north pole would move if free to do so, that is, **away from the magnetic north pole and towards the south**.

The molecular theory of magnetism

Wilhelm Weber first suggested this theory. He regarded each molecule of a magnetic substance to be a permanent magnet whether the whole were magnetised or not. In the unmagnetised state he considered them to be arranged in closed chains, as in figure 80a. When the substance is partly magnetised they are like (b). When fully magnetised or **saturated**, they take the form indicated in (c).

This theory also explains why the poles are situated at the ends, and when a magnet is broken, poles will appear at the breaks. The theory also shows why a single pole can never be obtained; each north pole must have a corresponding south pole. If a magnet is heated above a certain temperature, called its **Curie temperature**, it loses its magnetism. As the molecules are heated they vibrate more and more, and at this temperature they have enough energy to break out of line and fall back into their closed chains.

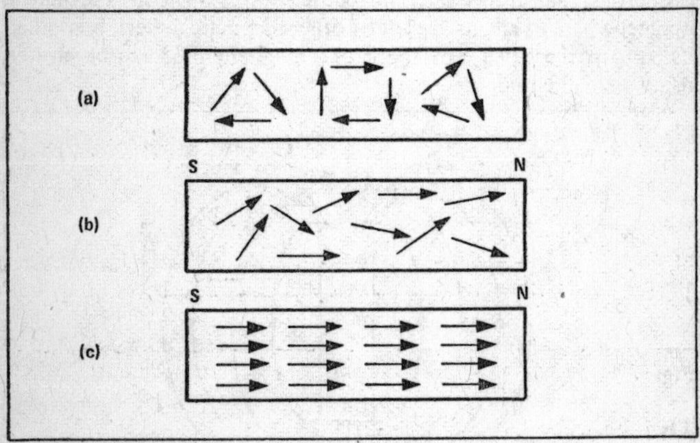

Figure 80. Molecular theory of magnetism

The magnetic properties of soft iron and steel

Steel is used to make **permanent magnets**. The molecules need a strong external field to move them into line, but once in line they remain there. **Soft iron**, on the other hand, is **very easily magnetised** but loses its magnetism easily as well. Iron nails and pins will become magnetised in the weak magnetic field at a distance from a magnet and will be attracted to it. This is called **magnetism by induction**. They will lose their magnetism when they are removed from the field of the permanent magnet.

Whether or not a magnetic material is magnetised can be investigated by testing whether or not one end of it is **repelled** by another magnet. Unmagnetised material which is temporarily magnetised by induction will be **attracted** by another magnet.

Methods of magnetisation

The most usual method is to put a metal bar in the strong magnetic field produced in a solenoid carrying an electric current (see page 147). Another method is to stroke the bar to be magnetised with a permanent magnet as shown in figure 81. The permanent magnet pulls open the closed chains and arranges the molecules in line.

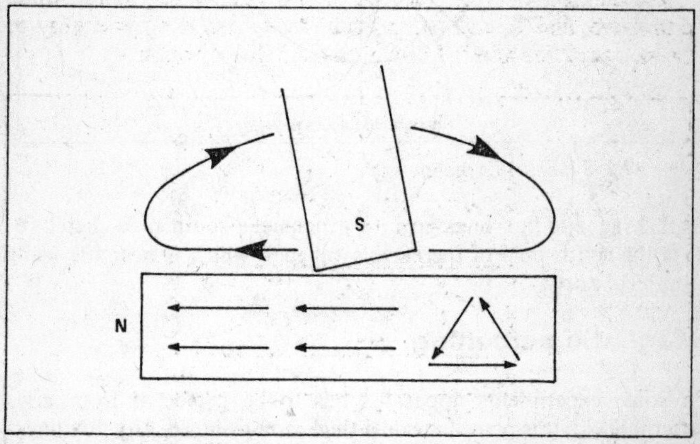

Figure 81. Magnetisation by stroking

The earth's magnetism

A compass needle points only approximately north-south. The angle between the direction in which the compass points and true north is called the **angle of declination**. This varies from place to place on the earth's surface. In England it is just over 10° west of north. If a bar magnet is freely suspended from its centre of gravity it will not only point north-south but will also point down at an angle of about 70° to the horizontal. This angle is called the **angle of dip**. The angle of dip also varies from place to place. Near the equator it is 0° and at the magnetic poles it is 90°. In its simplest form the earth's magnetic field can be likened to a large bar magnet situated at a slight angle to the geometrical axis (figure 82). The magnetic axis is slowly rotating about the geometric axis, once every 1 000 years, thus both the angles of declination and dip vary with time.

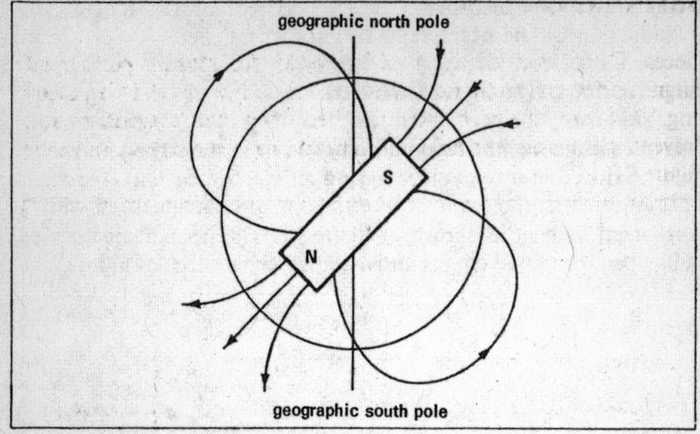

Figure 82. The earth's magnetic field

N.B. Magnetic flux lines go into a magnet's south pole therefore it is the south pole of the earth's 'magnet' which is near the geographical north.

Magnetic screening

In some experiments apparatus has to be protected from any external field. It is placed inside a thick iron cylinder. Any flux lines from outside will remain in the iron, leaving the centre free (figure 83).

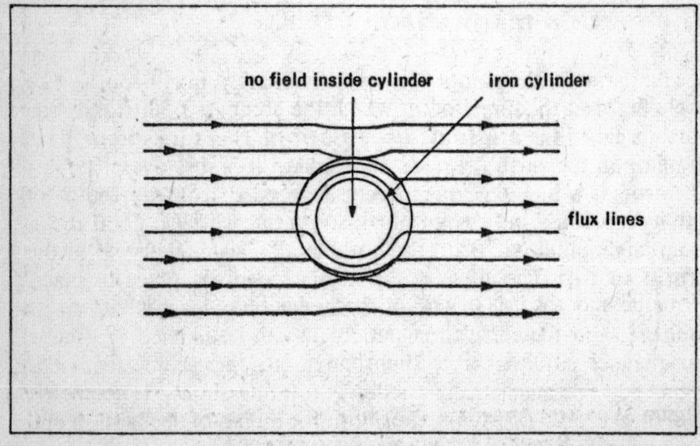

Figure 83. Magnetic screening

Soft iron keepers

Pieces of soft iron are used as **keepers** for storing permanent magnets. The magnetic field remains inside the metal as a closed ring. This helps the permanent magnets keep their magnetism and prevents their magnetic field interfering with other objects close by (figure 84).

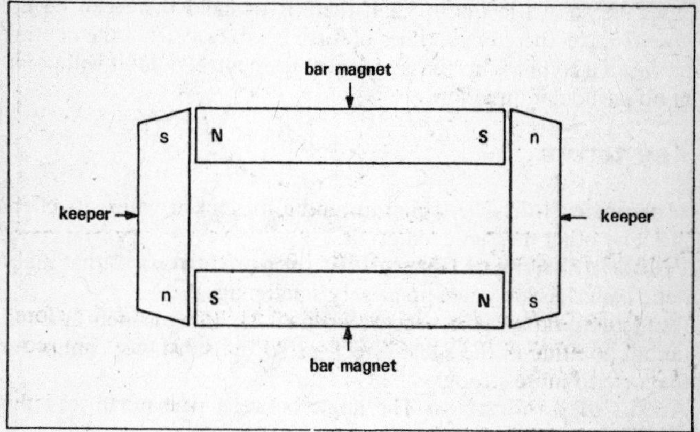

Figure 84. Use of soft iron keepers

Neutral points

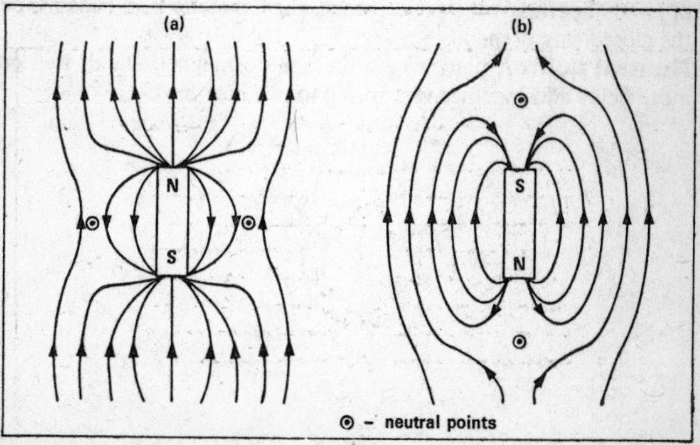

Figure 85. Neutral points: (a) N pole of bar magnet pointing north; (b) S pole of bar magnet pointing north

A magnetic field is **a vector quantity** (see page 29). It has both magnitude and direction. Therefore two magnetic fields that are equal in magnitude but opposite in direction will cancel each other out. The place where this happens is called a **neutral point**.

When the field created by a bar magnet is combined with the field created by the earth there are two neutral points. The two simplest cases are shown in figure 85. If there is no field there can be no lines of force, therefore all lines of force bend away from the neutral points. If a compass needle is placed at the neutral point it will point in no particular direction.

Key terms

Magnetic field The region around a magnet in which its effect is felt by other magnetic materials.
Weber's theory of permanent magnetism states that magnetic materials are made up of very small magnets.
Magnetic saturation occurs when all the little magnets in a rod are all pointing in the same direction, so the substance cannot be magnetised more strongly.
Angle of declination The angle between true north and the direction that a compass needle points.
Angle of dip The angle the earth's magnetic field makes with the horizontal.
Soft iron keepers Pieces of soft iron that are placed at the ends of pairs of permanent steel magnets. The magnetic field runs inside the closed ring of metal.
Neutral point A place where there is no magnetic field: two or more fields add together vectorially to produce zero.

Chapter 7
Electrostatics

Static (frictional) electricity

Plastic pens rubbed with a handkerchief will attract small pieces of paper and fluff. Nylon underclothes will often crackle and spark when taken off. Rubber balloons can be made to stick to the ceiling if they are first rubbed. These are some of the effects of static electricity, the most impressive effect being a thunderstorm. To produce all these effects two insulators must be rubbed together. (In the case of a thunderstorm, large clouds rub against the air surrounding them.)

As well as attracting small pieces of paper, two identical plastic pens if rubbed can be made to repel each other. Similarly two rubbed balloons will repel each other.

This phenomenon may be explained as follows. **There are two types of electric charge, positive** and **negative**. An uncharged body contains equal amounts of each. When two bodies are rubbed together some charge is pulled off one on to the other. **If two bodies with like charges are brought together they will repel each other. Unlike charges attract each other**. The unit of electric charge is called the **coulomb**.

A modern explanation

According to modern atomic theory, the atom consists of a small, **positively charged nucleus**, with a number of **negatively charged electrons** moving in rather complex paths around it. The nucleus contains a number of **positively charged protons**, and a number of **neutrons** which have no charge. Most of the mass of an atom is contained in the nucleus. The total number of protons and neutrons defines the **mass number** of the atom, while the total number of protons defines its **atomic number**. The magnitude of the positive charge on a proton is equal to the magnitude of the negative charge on an electron. Since a normal atom has an equal number of protons and electrons, it is electrically neutral (figure 86).

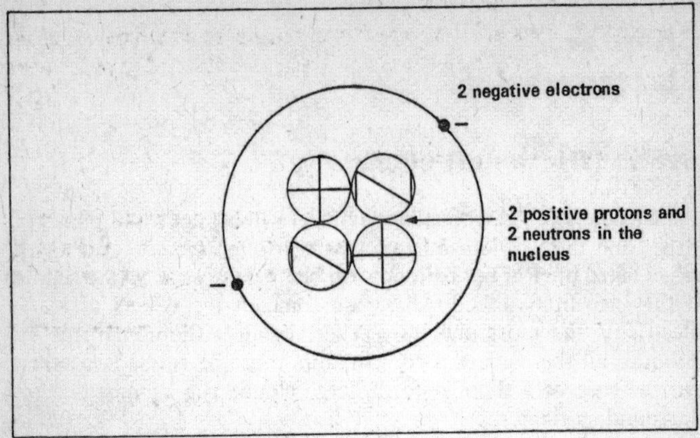

Figure 86. Structure of a helium atom

Materials that are **conductors** have one or more **electrons that are loosely connected to the nucleus** and can easily move through the material from one atom to another. The electrons of **insulating materials** are held fairly **firmly** to their nucleus. When two insulators are rubbed together some of the electrons on the surface of one are pulled on to the other. Thus one becomes positively charged and the other equally negatively charged.

Electrical potential

A charged body will try to get back to its neutral state by attracting opposite charges to itself. This force of attraction extends in the region around the body and is called its **electric field**. This movement of free charges from one point to another takes place only if the two points are at a different **electric potential**.

The magnitude and direction of the electric field between two points therefore depends on this **potential difference** between the two points. Potential difference is measured in **volts**. A positive charge produces a positive potential, a negative charge a negative potential. Electrons try to move from places of low potential to places of high potential. If a conductor is joined to places of different potential electrons will flow till the potentials are equal. Two points may be separated by an insulator, but if the potential difference between them is too great the insulator will break down and a spark will result. This is what happens in a lightning flash.

The earth as zero potential

The earth has an infinite capacity to absorb or produce electrons without itself becoming charged. Therefore it is always at zero potential.

Gold-leaf electroscope

This is an instrument (figure 87) for indicating electrical potential. If the leaf has an excess of charge, either positive or negative, it will be attracted to the earthed case and will diverge away from the central rod.

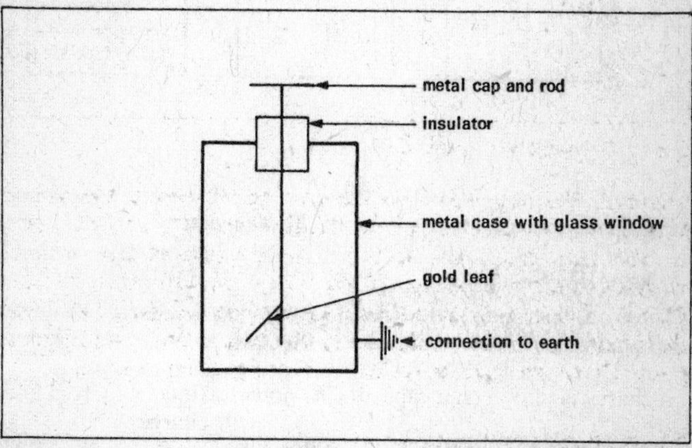

Figure 87. A gold-leaf electroscope

Charging a gold-leaf electroscope by induction

'By induction' means 'done at a distance', therefore in this instance the originally charged body does not touch the instrument.

The charged body is brought close to the top plate causing a separation of charge on this conductor. In this case, since the charged body is positively charged, it has a positive potential. Electrons on the leaf try to move as near as they can to it, so they move to the plate leaving a positive charge on the leaf. The leaf is therefore at a high potential and diverges.

The plate of the electroscope is now earthed by momentarily touching it with a finger. Electrons are attracted from the earth on to the plate owing to the presence of the positively charged body.

The charged body is removed and the leaf diverges because the negative charge distributes itself over the plate and leaf, giving it a high negative potential (figure 88).

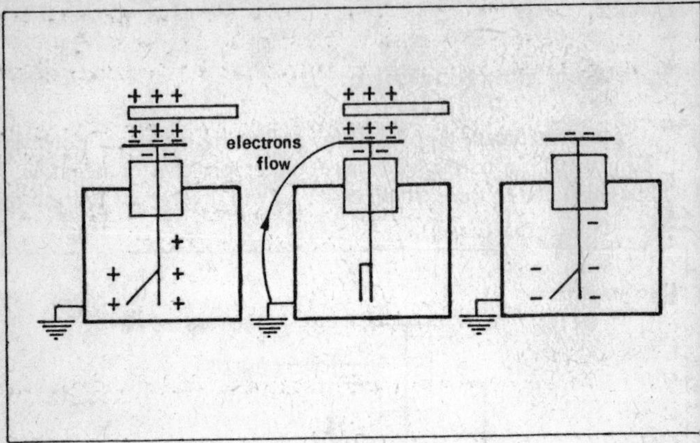

Figure 88. Charging an electroscope by induction

Capacitors

These are instruments designed to store electric charge. They were sometimes called **condensers** in the past.

If charge is placed on a capacitor its potential increases. It is found that the two are directly proportional, thus $\frac{\text{charge}}{\text{potential}}$ = a constant

The value of this constant is a measure of a capacitor's ability to hold charge and is called its **capacitance**. If the potential is measured in volts and the charge in coulombs then capacity will have units called **farads**:

$$\frac{\textbf{coulombs}}{\textbf{volts}} = \textbf{farads.}$$

Key terms

Insulator A substance that will not allow electricity to flow through it.
Conductor A substance that will allow electricity to flow through it.
Protons carry a positive charge and are found in the centre of the atom, the **nucleus**.

Electrons carry a negative charge. They orbit the nucleus.
Electric field A region around a charged body where its influence will be felt by other charged bodies.
Electric potential A quantity on which the strength and direction of an electric field depends. It is measured in volts (V).
Gold-leaf electroscope An instrument for measuring electric potential.
Induction The process by which bodies can be given a charge as a result of the influence of another charged body some distance away.
Capacitor A device which can store electric charge.
Farad The unit of measurement of capacitance. It indicates the ease with which a capacitor can store electric charge.

Example

Capacitance

What is the charge on a capacitor of capacitance $2\mu F$ if its potential is 3 000 volts?

Capacitance $= 2\mu F = 2 \times 10^{-6}$ F

$\therefore$ charge = capacitance × potential $= 2 \times 10^{-6} \times 3\,000$
$= 6 \times 10^{-3}$ coulomb, or 6 millicoulombs (mC).

Chapter 8
Current Electricity

In chapter 7 it was stated that a conductor consists of material in which there are some electrons which are free to move through the material. It was also stated, in the discussion of potential, that free negative charges will move from regions of low potential to regions of high potential. Thus if one end of a conductor is given a low potential, by supplying it with an excess of electrons, and the other end is given a high potential, by supplying it with an excess of positive charge, free electrons will flow along the conductor. Such a flow of electrons in a conductor constitutes an **electric current**.

There are various ways in which this **potential difference** (or p.d.) between the ends of a conductor can be produced. The first is by the action of the **electric cell**.

The electric cell

In an electric cell a chemical reaction produces a separation of charges. The negative (black) terminal has an excess of electrons. The positive (red) terminal has a deficiency of electrons. Thus there is a potential difference between the two terminals which is of the order of one or two volts.

The simple cell

This is made up of two plates of zinc (Zn) and copper (Cu) in a solution of sulphuric acid, H_2SO_4.

The sulphuric acid molecule, when dissolved in water, splits into **two hydrogen ions** and **one sulphate ion**. (An ion is an atom or molecule that has lost or gained one or more electrons.) The hydrogen ions, $2H^+$, are hydrogen atoms which have each lost an electron. These electrons are taken by the sulphate group to form a **sulphate ion**, SO_4^{--}, which has a double negative charge. The sulphuric acid is said to **dissociate** into positive and negative ions, according to the equation $H_2SO_4 \rightarrow 2H^+ + SO_4^{--}$.

The SO_4^{--} ion reacts with the zinc forming zinc sulphate, $ZnSO_4$, which is neutral and gives the two excess electrons to the zinc plate. The H^+ ion takes an electron from the copper plate and

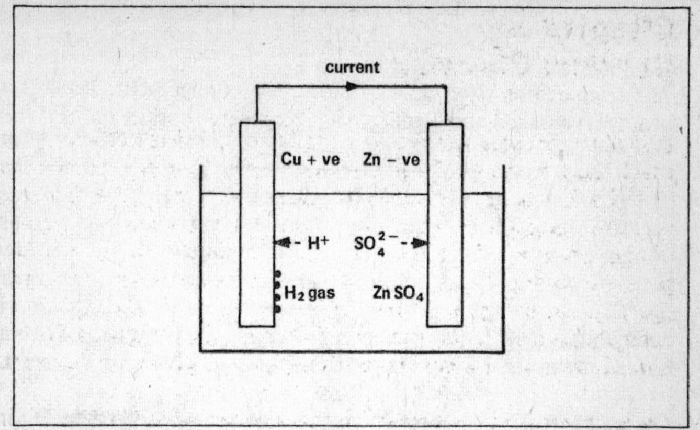

Figure 89. A simple cell

becomes neutral; neutral hydrogen is a gas and therefore bubbles form on the copper plate (figure 89). The zinc plate is therefore left with an excess of electrons: it is negatively charged; the copper plate is left with a deficiency of electrons: it is positively charged. The potential difference between the two plates before they are connected to a conductor is called the **electromotive force**, or e.m.f. of the cell. The e.m.f. is also measured in volts.

Faults of the simple cell

(1) **Polarisation** This is the presence of the hydrogen bubbles on the copper plate preventing further H^+ ions reaching the plate. A **back e.m.f.** is set up and the p.d. drops.

(2) **Local action** This occurs at the zinc plate because of impurities (other metals) in it. The two metals give rise to little cells on the surface of the electrode and the zinc gradually dissolves. Pure zinc is expensive but commercial zinc rubbed with mercury will form a zinc amalgam on the surface and the acid reacts with the zinc but not the mercury.

The Daniell cell

This overcomes the polarisation effect by introducing a second reaction that eliminates the formation of hydrogen gas (figure 90). The H^+ ions come through the porous pot and react with the copper sulphate, replacing the copper in the molecule and giving the copper a positive charge by taking two electrons away from the copper according to the equation

$$2H^{+} + CuSO_4 \rightarrow H_2SO_4 + Cu^{++}$$

The copper ions take electrons from the copper plate, leaving it positively charged and forming a further layer of copper on it. It has an e.m.f. of 1·08 volts.

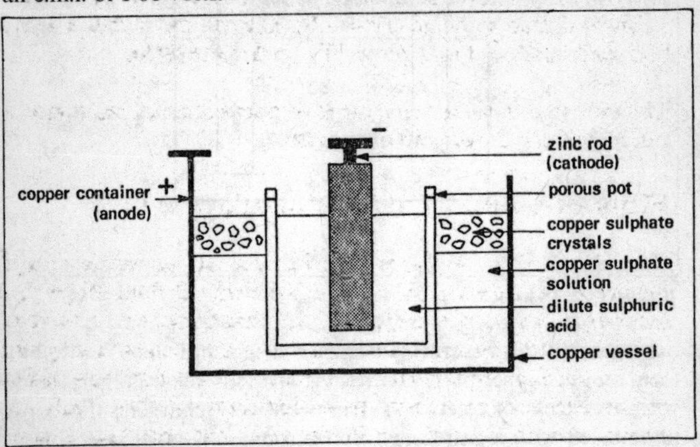

Figure 90. The Daniell cell

The Daniell cell has the disadvantage of having a **high internal resistance** (see pages 141–2), and also the chemicals need to be replenished after a time. These faults are overcome by the **secondary cell.**

Secondary cells

In secondary cells (also known as **accumulators** or **storage cells**), a chemical reaction is brought about by first passing a current through the cell. The chemical reaction is then reversed, and the cell produces an e.m.f. Currently the most common is the **lead-acid accumulator**, which starts off as two lead plates in dilute sulphuric acid. Current is passed through, hydrogen gas forms on one of the lead plates and lead peroxide on the other. The oxygen and the hydrogen come from the water so that during the charging the acid becomes stronger. After the initial charging the reverse reaction will not take place completely during discharge as the hydrogen bubbles will be lost and lead sulphate formed. However, water will also be formed, so the acid will become weaker and the lead peroxide will return to lead. This happens while current is being taken from the cell.

During a second and further recharging the hydrogen does not form bubbles on the plate but reacts with the lead sulphate forming sulphuric acid and lead.

A fully-charged lead accumulator has an e.m.f. of just over 2 volts. Its e.m.f. should never be allowed to fall below 1·8 volts as there is a danger of lead sulphate crystals forming on the plates. These are bad conductors and may physically buckle the plates.

This type of cell has the advantage of a low internal resistance and can be recharged over and over again.

Flow of charge through conductors

If an electrical conductor is joined across the terminals of a cell, the excess of electrons on the negative terminal flows through the conductor to the positive terminal. An electric current is produced. Though in a conductor the electrons flow from negative to positive it is conventional to consider the direction of the current as **a flow of positive charge from positive to negative**. This electric current cannot be seen, but in the right circumstances it can be detected in three ways.

(1) Heating effect

As the electrons move through a conductor they hit the atoms, or molecules, giving them energy, which causes them to vibrate more rapidly. As a result, the temperature of the conductor rises. This effect is made use of in filament lamps where the wire is heated to white hot, in electric fires and other type of heaters.

(2) Magnetic effect

When a current is flowing through a wire a **magnetic field** is set up around it. This effect is used in electromagnets, bells and buzzers, electric motors and transformers, and in a number of measuring instruments, e.g. the moving-coil galvanometer. Electromagnetic theory is explained on pages 146 ff.

(3) Chemical effect

An electric current will pass through certain liquids, called **electrolytes**, and at the same time produce a chemical change at the **electrodes**, where the charge enters and leaves the liquid. Thus water can be split into oxygen and hydrogen. This is also used for electro-plating.

Electrical units

The **ampère** is the unit of electric current. It is defined as that current which when passing through two infinitely long, straight parallel wires of negligible cross-section, in a vacuum, produces a force between the wires of **2×10^{-7} newtons per metre**.

This is a fundamental, defined S.I. unit, like the metre, the second and the kilogram. All the other electrical units can now be derived from these four.

When ammeters are calibrated at the National Physics Laboratory, the forces between coils of wires carrying currents can be calculated using the definition of electrical current stated above (obviously the Laboratory does not possess wires of infinite length). These forces can then be measured with highly accurate balances.

The **coulomb** is the unit of electric charge. It is defined as the charge passing any point in a circuit when a current of one ampère flows for one second.

Charge = current × time. Coulombs = amps × seconds.

The **volt** is the unit of electric potential difference and electromotive force. It can be considered, rather loosely, as the electrical force or pressure difference between two points.

More precisely, the unit of potential difference between two points is defined in terms of the work done in moving a charge of 1 coulomb from one point to the other.

$$\textbf{potential difference} = \frac{\textbf{work}}{\textbf{charge}} \qquad \textbf{volts} = \frac{\textbf{joules}}{\textbf{coulombs}}$$

This definition makes it possible to calculate the electrical energy used up when a current flows in a circuit:

energy = potential difference × charge
joules = volts × coulombs
or:
energy = potential difference × current × time
joules = volts × amps × seconds

Similarly it is possible to calculate the power developed in an electric circuit since power is the rate at which work is done.

$$\textbf{power} = \frac{\textbf{work}}{\textbf{time}} = \textbf{potential difference} \times \textbf{current}$$

$$\textbf{watts} = \frac{\textbf{joules}}{\textbf{seconds}} = \textbf{volts} \times \textbf{amps}$$

Ohm's law

This states that **the current flowing through a conductor is directly proportional to the p.d. across its ends provided physical conditions such as temperature remain constant.**

Thus: $\frac{\text{p.d.}}{\text{current}}$ = constant, which is called the **resistance** of the conductor. The units of resistance will be volts/amps and these are given the special name **ohms**.

$$\frac{\text{volts}}{\text{amps}} = \text{ohms} \quad \text{or} \quad \text{volts} = \text{amps} \times \text{ohms} \quad \text{or} \quad \text{amps} = \frac{\text{volts}}{\text{ohms}}$$

Experimental verification of Ohm's law

Connect three accumulators, a voltmeter, an ammeter and a conductor having some resistance as indicated in figure 91, first with three cells, then with two and then with one, noting the values of the voltmeter, V, and of the ammeter, I, each time. Tabulate V, I and V/I. The latter should be constant, being the value of the resistance, R, in ohms.

As a precaution the resistance should be immersed in a beaker of water to keep its temperature constant.

Devices in which resistance is not constant

The constant value of R is very useful in calculations relating to small-current electrical circuits, which produce little heat. However, in components such as light bulbs the temperature of the wire increases considerably and R increases. As a model, one can picture the free electrons moving through the lattice of atoms, and as the metal gets hotter the atoms vibrate more rapidly about their lattice

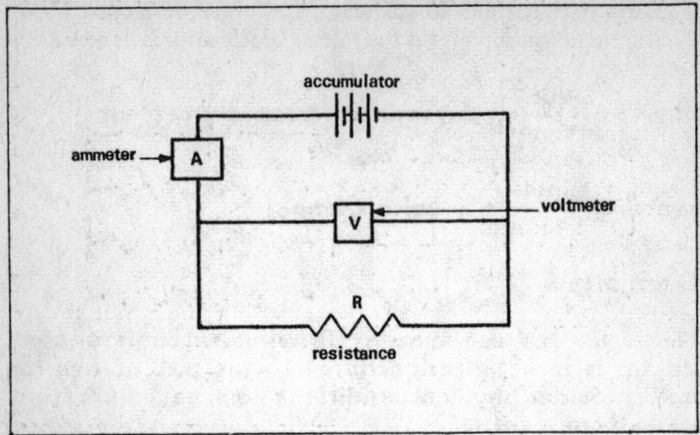

Figure 91. Circuit for verifying Ohm's law

point and so the electrons find it more difficult to move. Metals thus have a **positive temperature coefficient of resistance**: that is, their resistance increases with temperature.

An increasing number of electrical components made up of **semiconductors** is now being manufactured. The **thermistor** has a negative temperature coefficient of resistance: that is, its resistance decreases with temperature. What is happening here is that at higher temperatures more electrons are being set free to carry the charge. Another example is the photo-resistor; in this, light falling on the device releases electrons, thus allowing a current to flow. When the light source is removed the device takes on a very high resistance again.

Resistances in series and in parallel

(1) Resistances in series

The total resistance of the circuit is simply the sum of the individual resistances. If v_1 is the p.d. across resistance r_1, v_2 across r_2 etc. then the total potential drop across them all, $V = v_1 + v_2 + v_3$ (figure 92a).

The same current, I, passes through them all. Therefore if R is the total effective resistance:

$$R = \frac{V}{I} = \frac{v_1 + v_2 + v_3}{I} = \frac{v_1}{I} + \frac{v_2}{I} + \frac{v_3}{I}$$

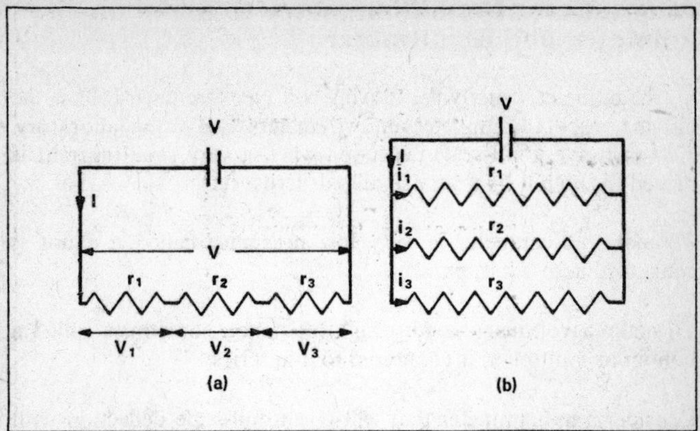

Figure 92. (a) Resistances in series; (b) resistances in parallel

But $\frac{v_1}{I} = r_1 \quad \frac{v_2}{I} = r_2$ etc.

Thus $R = r_1 + r_2 + r_3$

(2) Resistances in parallel

The total current I splits into i_1 through r_1, i_2 through r_2 etc. so that $I = i_1 + i_2 + i_3$ (figure 92b).

The p.d. across all the resistances is V.

Thus, if R is the effective resistance:

$$R = \frac{V}{I} = \frac{V}{i_1 + i_2 + i_3}$$

Turn each side of the equation upside down:

$$\frac{I}{V} = \frac{i_1 + i_2 + i_3}{V} = \frac{i_1}{V} + \frac{i_2}{V} + \frac{i_3}{V}.$$

$$\therefore \quad \frac{1}{R} = \frac{1}{r_1} + \frac{1}{r_2} + \frac{1}{r_3}.$$

The reciprocal of the total resistance is the sum of the reciprocals of the individual resistances.

Adaptation of a milliammeter to make an ammeter and a voltmeter

A milliammeter, usually the moving coil type (see page 150), is the moving part of the ammeters and voltmeters used in the laboratory. This will give a full-scale deflection when a very small current is passed through it by a very small p.d. across it.

To make an ammeter, a very low resistance called a shunt is connected across it in parallel.

To make a voltmeter, a very high resistance, sometimes called a bobbin or multiplier, is connected to it in series.

Consider a milliammeter that will give a full-scale deflection with i amps flowing through it, and v volts across it. It will then have a resistance r ohms such that $r = v/i$.

If it is required to produce an ammeter to give a full-scale deflection with I amps, a resistance R ohms must shunt most of the current round the instrument (figure 93a).

$(I-i)$ amps must go through the shunt with a p.d. across it of v volts, therefore it must have a value of $\frac{v}{(I-i)}$ ohms, or $\frac{i.r}{(I-i)}$ ohms.

If the voltmeter is required to give a full-scale deflection with V volts, a resistance R ohms must be placed in series so that most of the potential drop takes place across R.

$(V-v)$ volts is the p.d. across R and the current is i amps, thus R has a value of:

$\frac{(V-v)}{i}$ ohms, or $\frac{V}{i} - r$ ohms (figure 93b).

Resistivity

The resistance of a piece of wire depends on its length, its area of cross-section and the material from which it is made. The resistance, R ohms, increases with length, l m, and decreases with bigger cross-sectional area, A m^2.

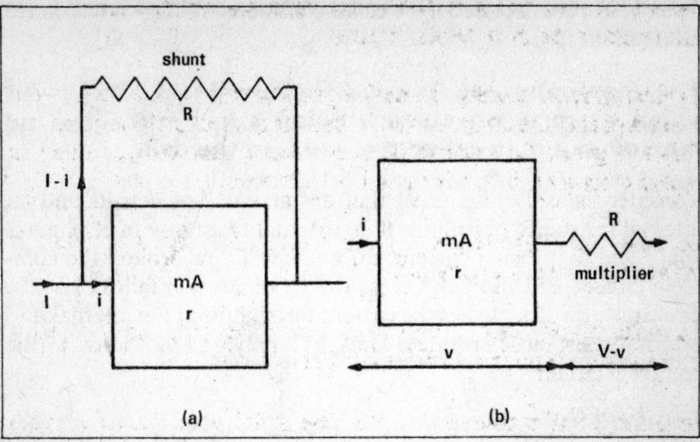

Figure 93. Adaptation of a milliammeter: (a) as an ammeter; (b) as a voltmeter

$\therefore$ R is proportional to $\frac{l}{A}$

or $R = \rho \frac{l}{A}$.

Where ρ is a constant it is called the **resistivity** of the wire. It can be considered as the resistance from one face to the opposite face of a cubic metre of the substance. **ρ has units of ohms × metres.**

Temperature coefficient of resistance

One condition that Ohm's law should be obeyed is that the temperature of the resistance is constant. If the temperature increases the resistance increases. The formula for the increase is: $R_\theta = R_0(1+\alpha\theta)$. R_θ is the resistance at θ°C and R_0 is the resistance at 0°C. α is the **temperature coefficient of resistance.**

This is the theory of the **resistance thermometer**. If α is known, it is possible, by measuring the resistance of a coil of wire, to calculate its temperature, provided its resistance at 0°C is known.

E.m.f. and internal resistance of cells

It is found that the p.d. across the terminals of a cell decreases as larger values of current are taken from it. This decrease can be explained and calculated if the cell is considered to have a resistance inside itself.

The maximum p.d. across the terminals will occur when there is no current being taken from it. This maximum p.d. is called the e.m.f. of the cell.

Consider the cell in figure 94 that has an e.m.f. of E volts and an internal resistance of r ohms. If an external resistance of R ohms is connected to it then a current of I amps will flow around the complete circuit, and the p.d. across the terminals will fall to V volts. The e.m.f., E volts, pushes the current through both the internal and external resistances in series. Thus, by applying Ohm's law to the complete circuit,

$$E = I(R+r)$$

V volts is the p.d. across the external resistance.
Thus: $V = IR$
The 'lost volts', the apparent decrease in the voltage of the cell, is $E-V$. This pushes the current through the cell resistance.

Thus: $E-V = Ir$

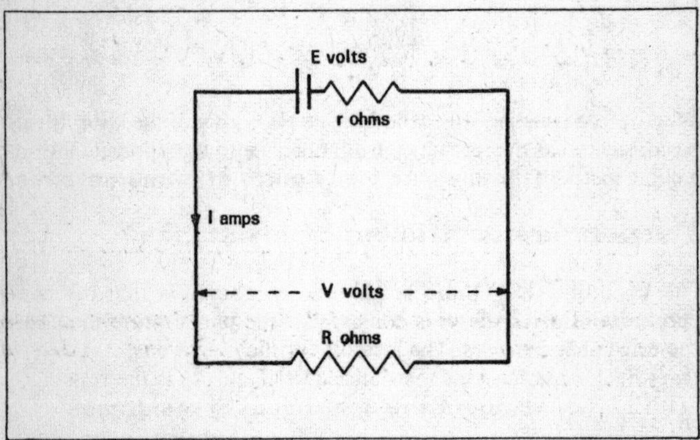

Figure 94. Complete circuit showing internal resistance of cell

The potentiometer for comparing the e.m.f.s of two cells

The definition of e.m.f. as the p.d. across the terminals of a cell with **no current flowing** means that an ordinary voltmeter such as a moving coil instrument cannot be used to measure the e.m.f. of a cell because it takes a certain amount of current to work it. The **potentiometer** is an instrument which takes no current from the source of p.d. that it is measuring.

The main part of the instrument is a **uniform-resistance wire**, *AB* in figure 95. A current, *I* amps, is passed through it by a battery causing a uniform **potential drop** along the wire. The cell under investigation, C_1, has one terminal connected to *A* and the other to a centre zero galvanometer. The other side of the galvanometer is connected to the movable contact *D*. The position of *D* is found where there is no deflection of the galvanometer. In this position the p.d. along the resistance wire from *A* to *C* must balance the p.d. across the cell, and as no current is flowing from the cell this is its e.m.f., E_1 volts.

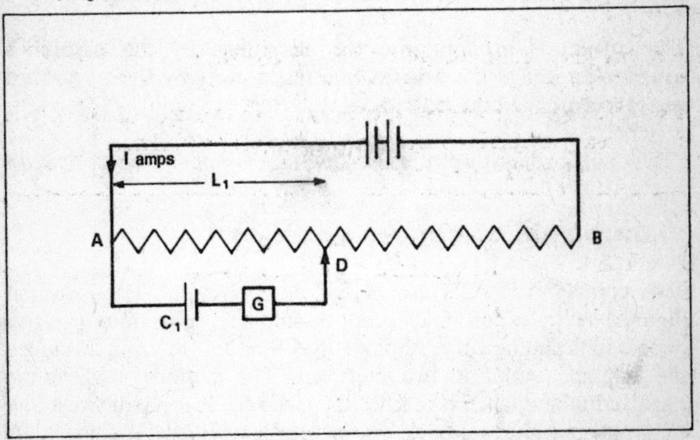

Figure 95. The potentiometer

Let the wire *AB* have a resistance of *R* ohms per centimetre. Let the battery produce a current of *I* amps through *AB*. Then as *AD* is of length L_1 the p.d. across it is $I.R.L_1$ volts.

$$E_1 = I.R.L_1$$

If cell C_1 is replaced by a second, C_2, and a balance obtained with a length L_2,

$$E_2 = I.R.L_2$$

Dividing one equation by the other, *I* and *R* cancel.

$$\frac{E_1}{E_2} = \frac{L_1}{L_2}$$

The battery must produce a p.d. across AB that is bigger than the e.m.f. of either cell, but it does not enter into the calculation. Note that the cells and the battery have the same sign terminal connected to A.

Usually one of the cells is a standard, the e.m.f. of it being known accurately, e.g. a Weston cell of e.m.f. 1·018 volts, hence the e.m.f. of the other cell can be calculated.

Electrolysis

Most liquids which conduct electricity are split up chemically by the current. The phenomenon is known as **electrolysis** and the liquids are called **electrolytes**.

The current is brought into the electrolyte by the **positive electrode**, called the **anode**, and taken away by the **negative electrode**, called the **cathode**.

The vessel in which electrolysis takes place is called a **voltameter**.

Electrolysis of copper sulphate

Two copper electrodes are placed in copper sulphate solution (figure 96). In solution, the copper sulphate splits into positive copper ions and negative sulphate ions. The copper atom has given the sulphate molecule two electrons. The cathode attracts the positive ion and when it reaches the electrode it receives two more electrons and changes back into neutral copper metal again, thus forming a layer of fresh copper on the electrode.

The sulphate ion is attracted to the anode, gives up its two electrons to the anode, but as neutral sulphate is unstable it attacks the copper and pulls one more atom into solution.

The net result is a movement of electrons from cathode to anode, i.e. a conventional positive current runs from anode to cathode, and copper is removed from the anode and deposited on the cathode.

Faraday's laws of electrolysis

(1) **The mass of a substance liberated is proportional to the quantity of electric charge passed.** $M \propto Q$. But $Q = It$.

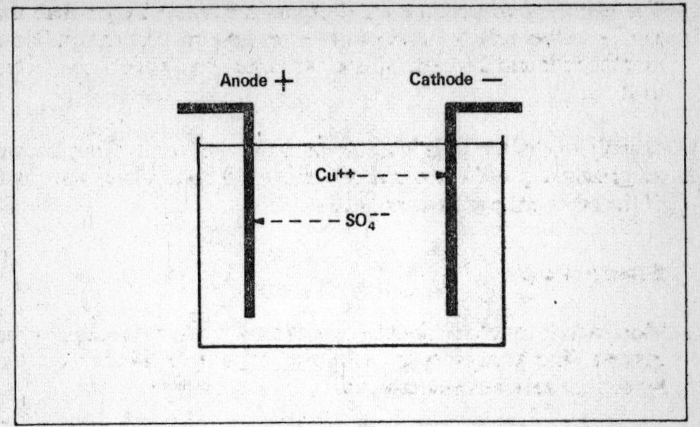

Figure 96. Electrolysis of copper sulphate in a copper voltameter

Thus $M \propto It$ or $M = zIt$, where z is a constant called the **electrochemical equivalent** of the substance. z is the mass liberated when one coulomb passes and thus has units of kg/C.

(2) **When the same quantity of electric charge passes through different electrolytes the masses of substance liberated are in the ratio of their chemical equivalent weights.**

Electrolysis of water

For this process, platinum electrodes are used and the voltameter is designed so that the gases produced at the electrodes can be collected and their volumes measured (figure 97). Pure water is a good insulator: the water molecules (H_2O) hardly split up into ions. If, however, a little sulphuric acid is added the water becomes an electrolyte. The water molecules can be considered to be split into positive hydrogen ions and negative hydroxide ions, H^+ and OH^-.

The H^+ ion is attracted to the cathode where it receives an electron and turns into neutral hydrogen gas.

The OH^- ion is attracted to the anode to which it gives its spare electron. Two of these neutral OH molecules combine to produce a molecule of water and one atom of oxygen, according to the equation $2OH \rightarrow H_2O + O$.

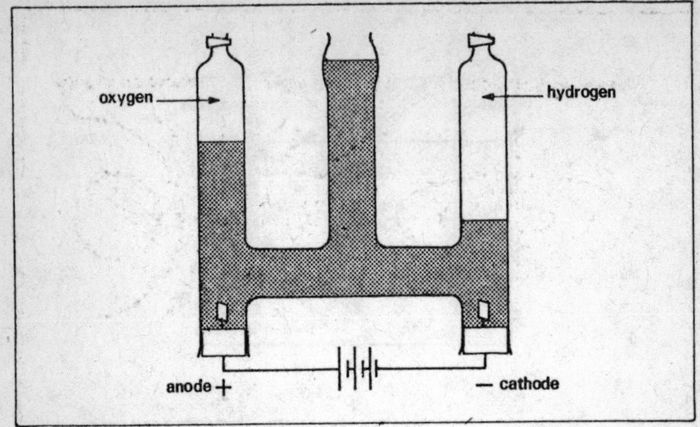

Figure 97. Electrolysis of water

Note that to produce one atom of oxygen requires two ions, OH^-, and thus two atoms of hydrogen will be formed at the cathode. Thus the volume of hydrogen formed is twice that of oxygen.

The net result is that electrical energy is changing water into hydrogen and oxygen.

If more active metals are used instead of platinum for the electrodes, the oxygen is likely to react with the anode forming an oxide and oxygen gas will not be formed.

Magnetic effect of an electric current

The fact that a current flowing in a wire has a magnetic field associated with it was first discovered by Oersted in 1819, when he noticed that a compass needle was deflected when placed close to the wire. Further investigation showed that **the magnetic field created by a current passing through a long straight wire is circular** (figure 98a). The direction of the field is given by the **corkscrew rule**. If one tries to drive a screw in the direction of the current, the direction of turning will give the direction of the field. The magnetic field created by a current flowing in a **circular wire loop** is shown in figure 98b. Note that the field has the same direction everywhere inside the loop. Use the corkscrew rule to verify this.

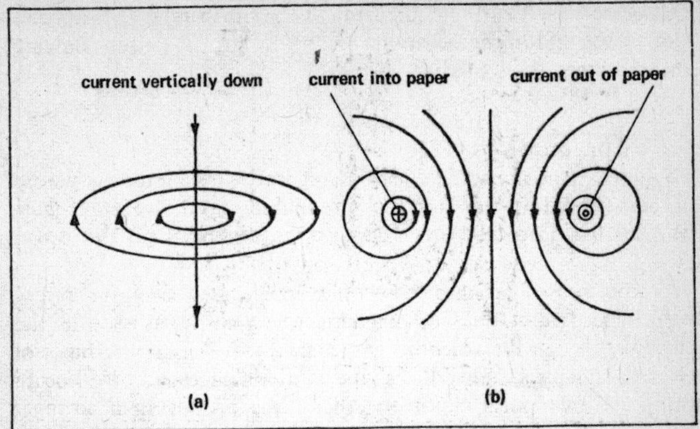

Figure 98. (*a*) *Field created by wire;* (*b*) *field created by loop*

Magnetic field created by a current flowing through a solenoid

A solenoid is a long thin coil consisting of many turns of wire. The field produced by each turn of wire adds to the next, thus the total field in the centre of the coil is very strong. The shape of the field outside the solenoid is identical to that of a bar magnet (figure 99a).

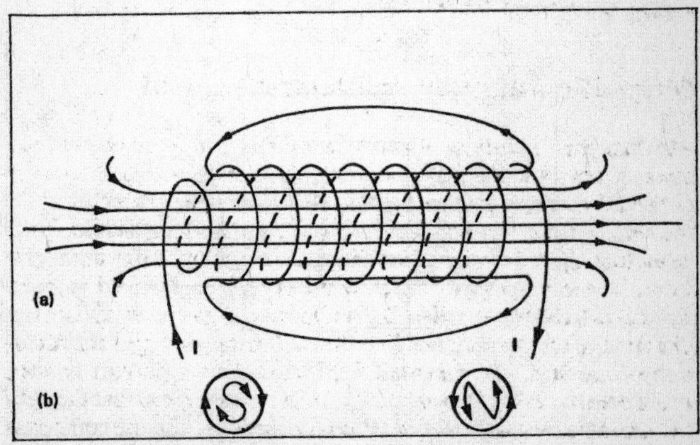

Figure 99. (*a*) *Magnetic field around a solenoid;* (*b*) *polarity of the ends*

To decide which end is equivalent to the north pole of a magnet look at the end of the solenoid. The north has the current flowing anti-clockwise (figure 99b).

Soft iron cores

To increase the magnetic effect, a soft iron core should be placed in the solenoid. In this case the wire making up the solenoid must be insulated. **The current does not pass through the core.**

Soft iron is used because it becomes magnetised easily in a very weak field. It also loses its magnetism very easily, as soon as the current through the solenoid is switched off. This is the basis of electromagnets. Usually the shape is modified to a horseshoe to bring the two poles closer together, thus producing a stronger external field.

The electric bell

The electromagnet is used in bells and buzzers (figure 100). When the circuit is completed by pushing the bell-push a current flows through the electromagnet and the magnetic field produced attracts the piece of soft iron attached to the hammer. Thus the bell is struck. But in doing this the contact, *X*, is broken, stopping the flow of current.

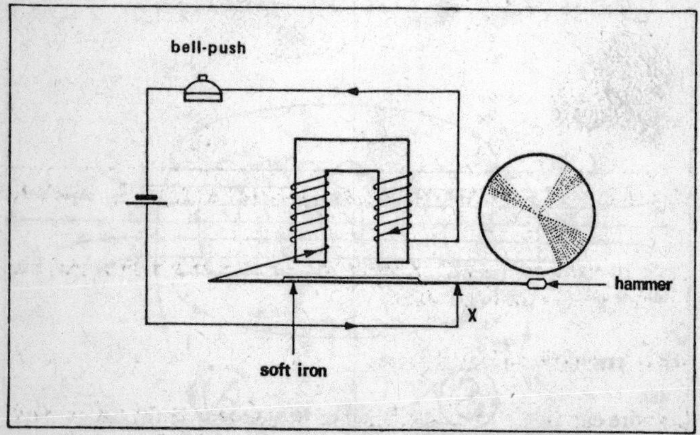

Figure 100. The electric bell

This causes the electromagnet to lose its magnetism so the hammer springs back to its original position. The contact X is again made, the current flows and the cycle starts again. This is called a 'make-and-break' circuit.

Moving iron ammeters

(1) Attraction type

A piece of soft iron is pivoted on a pointer that moves over a scale. A hair-spring keeps the iron just outside a small solenoid when no current is flowing through it. With a flow of current through the solenoid the iron is attracted into it. The greater the current the further it is attracted, moving the pointer over the scale (figure 101a).

(2) Repulsion type

Here there are two rods of soft iron, one fixed and the other attached to a pointer. When a current flows through the coil both become magnetised and repel each other (figure 101b).

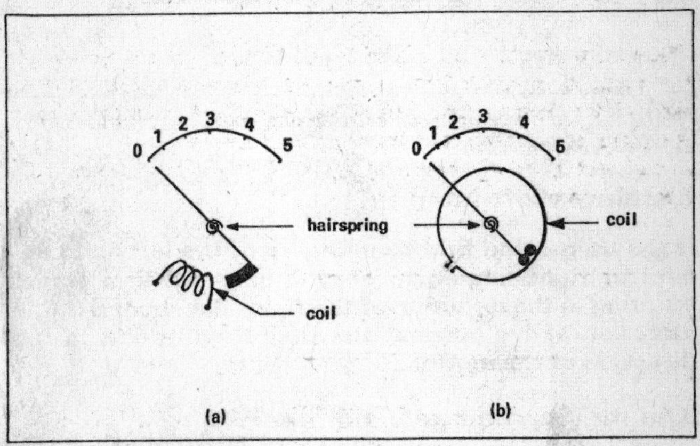

Figure 101. Moving iron ammeters: (a) attraction type; (b) repulsion type

Both of these types will measure direct and alternating current. Neither gives a uniform scale.

The motor effect

If a wire carrying a **current** lies in a **magnetic field** the wire experiences a **force**. A simple demonstration of this effect is indicated in figure 102a. A thin piece of aluminium foil carries the current between the poles of a permanent magnet. With a current flowing the foil bows inward. If the current or the field is reversed the direction of the force is reversed. All three quantities are at right angles to each other and their directions given by **Fleming's left-hand rule** (figure 102b).

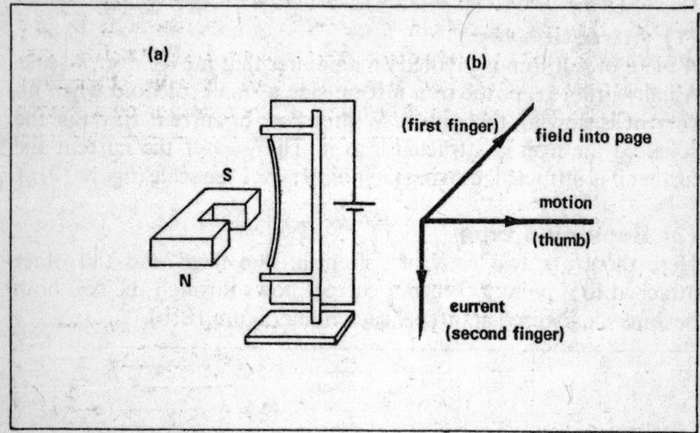

Figure 102. (a) Force on a current-carrying conductor; (b) Fleming's left-hand rule

Fleming's left-hand rule

If the thumb and first two fingers of the left hand are held at right angles to each other, the first finger pointing in the direction of the field, the second in the direction of the current, the thumb will point in the direction of the motion.

The moving coil galvanometer

This is the basis of many electrical measuring instruments and can be made both sensitive and accurate.

Basically it consists of a rectangular coil of wire suspended in a magnetic field so that it can rotate. When a current passes through the coil it experiences a **couple** that rotates it against some **restoring couple**. The final equilibrium position will depend on the strength of the current (figure 103a).

Practical details of the instrument should be noted. The pole pieces of the magnet are made concave and a **soft iron core** is fixed in the centre. This makes the magnetic field **radial** instead of straight across (figure 103b). The force on the sides of the coil will always be at right angles to the coil giving maximum **torque**. In a straight field the couple would vary, and thus give a non-uniform scale.

The coil is wound on an **aluminium 'former'**. As the coil moves through the magnetic field a current will be induced in the aluminium, called an **eddy current** (see page 156) that will oppose the motion (see Lenz's law, page 153). This provides **damping** which makes the coil quickly come to rest in its equilibrium

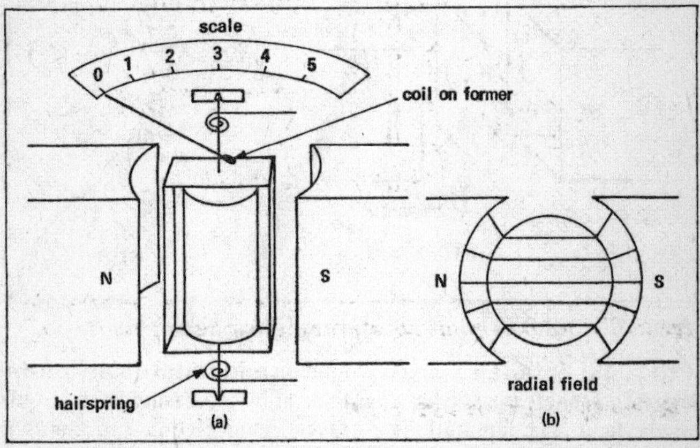

Figure 103. (a) Moving coil galvanometer; (b) radial field

position. Most instruments have the coil mounted on **jewelled bearings** and the restoring couple produced by **hair-springs**. The springs also act as leads to the coil. In more sensitive instruments the coil is suspended by a phosphor-bronze strip. A counter-balanced pointer fixed to the coil moves over a scale. The bigger the pointer the more sensitive will be the readings. For most accurate work the pointer is replaced by a mirror which reflects a beam of light on to a scale.

The D.C. electric motor

A coil will rotate in a magnetic field until the plane of the coil is perpendicular to the direction of the field. Here there is no couple acting on it. The forces on the two sides of the coil are in line (figure 104a). To produce continued rotation the current must be switched off as the coil goes through this position and then reversed.

This switching off and reversing is done by a **split-ring commutator** (figure 104b). The connections to the coil are made by two carbon **'brushes'** that press on to the two half-rings which rotate with the coil. Note that the half-rings are arranged so that the current is reversed when the coil is **at right angles to the magnetic field**.

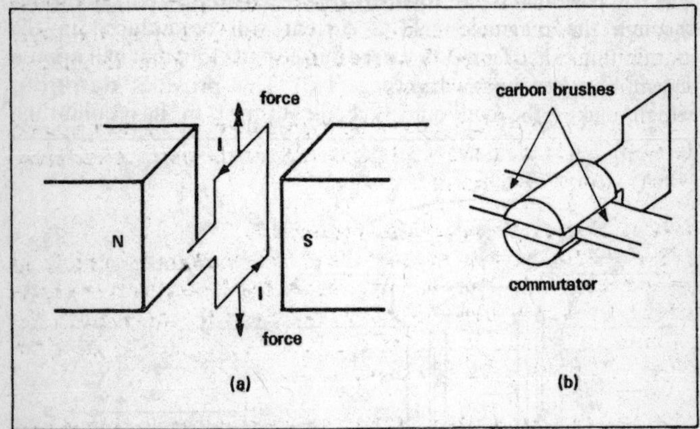

Figure 104. (a) D.C. motor; (b) split-ring commutator

Practical details The coil is wound on a **laminated soft iron core**. This increases the force on the coil by increasing the strength of the field. The laminations, or layers, reduce eddy currents (see page 156). The rotating part of the motor is called the **armature**. With the soft iron core the armature is fairly heavy and this produces smoother running.

If only one coil is used the motor is not self-starting, the coil may come to rest with the brushes across the break of the split-ring and the armature will have to be moved to start it. To prevent this happening and to give smoother running two further coils are added at 60° to each other. The split-ring commutator will have six parts and the brushes will always be connected to at least one coil.

In a D.C. motor the magnet may be a permanent one or an electromagnet.

For an **A.C. motor** the magnetic field must be produced by an **electromagnet** connected to the **same A.C. source**. The fields produced by the electromagnet and the armature will both be changing in step with each other and so the effect will be the same as if both were steady fields.

Electromagnetic induction

If a bar magnet is pushed into a coil of wire the free ends of which are connected to a galvanometer, the latter will show a deflection. The moving magnet has created an **induced e.m.f.** in the wire causing a current to flow. If the magnet is withdrawn the induced e.m.f. produces a current in the opposite direction. If the magnet is pushed in faster the induced e.m.f. is greater, although the larger current that is produced flows for a shorter time. When the magnet is held at rest no current flows.

The same effect always occurs whenever there is relative motion between a wire and a magnet (figure 105a). The effect is increased when a coil of many turns is used.

Laws of electromagnetic induction

(1) Faraday's law states that the induced e.m.f. is proportional to the rate at which the lines of the magnetic field are cut. Therefore the faster the movement, the stronger the magnetic field; and the greater the number of coils, the bigger will be the e.m.f. produced.

(2) Lenz's law states that the direction of the induced e.m.f. is such as to oppose the movement causing it. This follows from the law of conservation of energy. To produce electrical energy mechanical energy must be used up.

(3) Fleming's right-hand rule gives the directions of

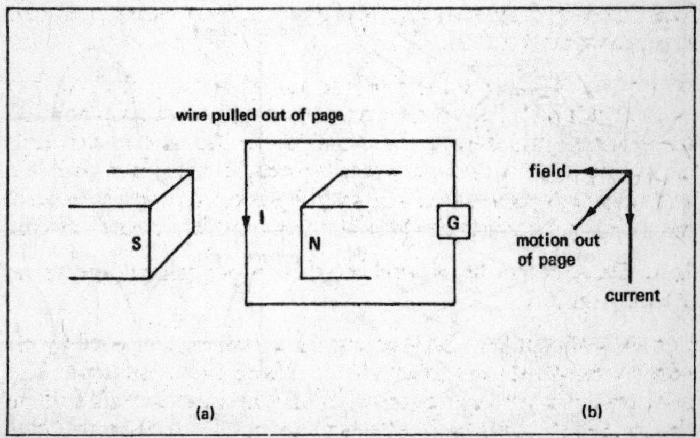

Figure 105. (a) Electromagnetic induction; (b) Fleming's right-hand rule

the field, motion and the current produced. The thumb and first two fingers of the right hand are held at right angles to each other. If the first finger points in the direction of the magnetic field and the thumb points in the direction of the motion, the second finger will point in the direction of the induced current (figure 105b).

Note Fleming's right-hand rule: magnetic field and motion produce current. Fleming's left-hand rule: magnetic field and current produce motion.

The current direction is that of the conventional positive charge that flows from positive to negative, not the flow of electrons. The magnetic field direction is conventionally from north to south.

Alternating- and direct-current generators

A coil of wire is rotated in a magnetic field. As the coil rotates it cuts the lines of the magnetic field and an e.m.f. is induced in the coil. When the coil is in position *A* the rate of cutting the field lines is a maximum, giving maximum e.m.f. The induced e.m.f. decreases till it is zero in position *B* where the sides of the coil are moving

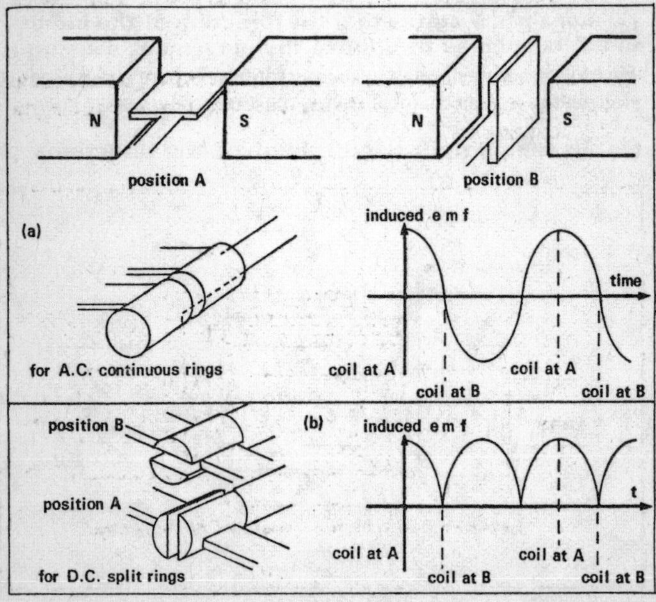

Figure 106. (a) A.C. generator; (b) D.C. generator

along the lines of the field. After passing this position the induced e.m.f. is in the reverse direction in the coil. The induced e.m.f. is first in one direction and then in the other during each rotation of the coil, i.e. alternating. The connections from the moving coil are through a commutator and fixed brushes. For A.C. each brush is connected to the same end of the coil for the complete rotation by a continuous ring (figure 106a).

For D.C. the connections are reversed when the coil is perpendicular to the field, i.e. when it is in position *B* (figure 106b). The direction of the induced e.m.f. supplied to the external circuit can be shown graphically, as in the figure.

Transformers

It was seen on page 153 that a magnet being pushed in and out of a coil of wire will produce an alternating e.m.f. in the coil. The same effect will be produced if the moving magnet is replaced by a stationary coil of wire through which a current is switched on and off.

The effect is increased considerably if both coils are wound round a soft iron core.

If instead of a direct current being switched on and off, an alternating current is applied, the same effect will be obtained. This is the basis of the transformer.

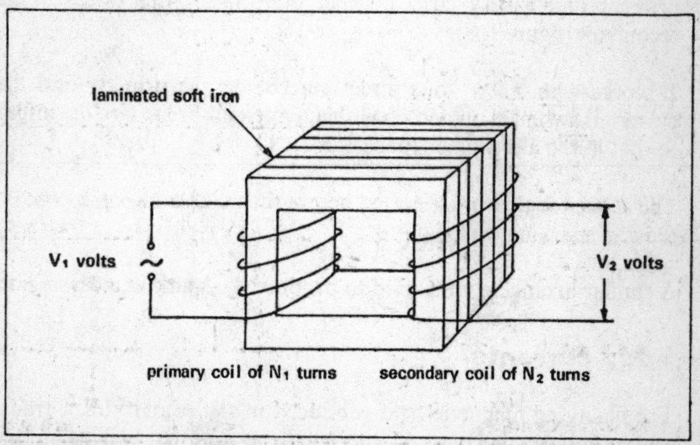

Figure 107. A transformer

An alternating potential, V_1 volts, is applied to the primary coil of N_1 turns. This produces an alternating magnetic field in the soft iron, which in turn induces an alternating potential, V_2 volts, of the same frequency in the secondary coil of N_2 turns (figure 107).

If the transformer is 100 per cent efficient, then:

$$\frac{V_1}{V_2} = \frac{N_1}{N_2}$$

In fact V_2 is a little less.

The coils are wound round laminated soft iron. Each lamination is electrically insulated to minimise eddy currents.

Thus a transformer can be used to increase or decrease the potential of an alternating voltage.

Note that in the step-up transformer the voltage is increased but the current will be correspondingly decreased. If the transformer were 100 per cent efficient the power input would equal the power output: $V_1 I_1 = V_2 I_2$.

Induction coil

This is an instrument for producing very high voltages. It is basically a step-up transformer with a few turns of thick wire as the primary and a very large number of turns of thin wire as the secondary (figure 108).

It works with a few volts direct current on the primary and the current is switched on and off with a make-and-break circuit similar to that in the electric bell (page 148).

The capacitor that is connected across the make-and-break contact reduces sparking across it.

A similar arrangement is used to produce the spark in a car engine.

Eddy currents

The changing magnetic field produced in the primary of a transformer induces currents in the soft iron core as well as in the secondary coil. These currents will produce heating in the core and

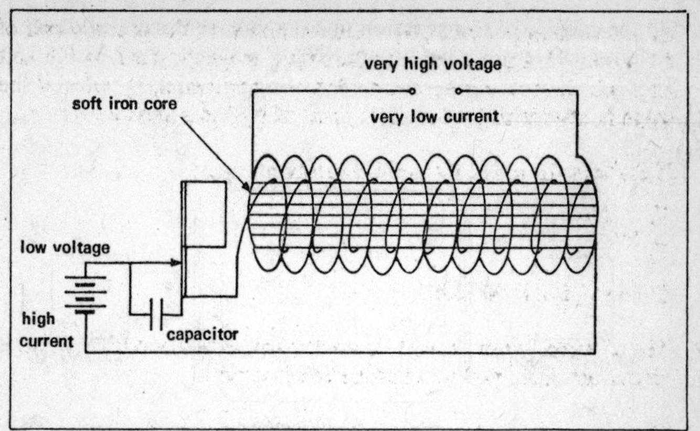

Figure 108. Induction coil

make the transformer inefficient. To reduce these **eddy currents** the core is not made in a solid mass but in strips or laminars that are painted with an insulating material.

In most cases, eddy currents are a disadvantage and have to be minimised. However, there are two practical uses.

One is for **damping**, the induced current being in such a direction as to oppose the movement causing it. A conductor moving through a magnetic field is brought quickly to a standstill.

The second is in an induction furnace. The eddy currents in a piece of metal at the centre of a coil cause it to heat so much that it melts. Although only small amounts of metal can be heated at one time, it is very suitable when high purity is required.

Power transmission

One of the main uses of transformers is for stepping up the voltage from the generators for power transmission, and for stepping it down again for use in factories and homes.

The transmission lines are made to have as low a resistance as possible. However, the power loss in them is I^2R and so it is even more important to have as low a current as possible. Thus for a certain power output the lower the current and the higher the voltage the better. But high voltages are dangerous and require high-quality insulators to handle them.

At the end of the transmission line the voltage is stepped down to a convenient level for use (figure 109).

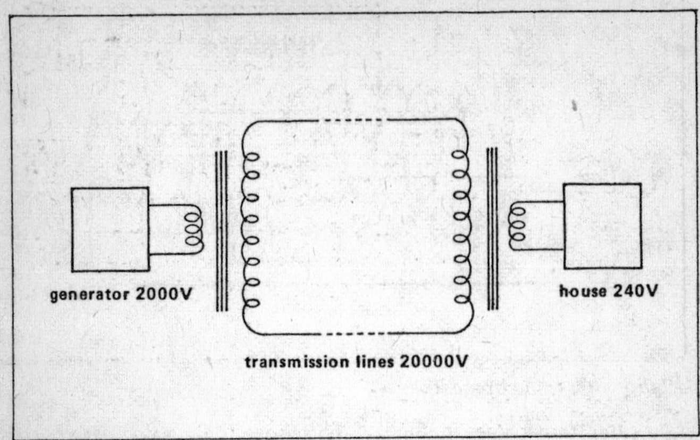

Figure 109. Power transmission

House wiring

The supply which electricity boards provide for houses is usually of 240 volts. It is alternating at 50 hertz. The blue lead is neutral, the brown is live. This means that the brown lead has the alternating voltage from positive to negative and back 50 times a second relative to the blue lead. All the appliances in the house are designed to work at 240 volts and must be joined in parallel.

Fuses Just as a chain has its weakest link, where it will break if overloaded, so an electric circuit will break at its weakest point if too large a current is taken from it. The fuse is designed to be that weakest point.

Electricity boards have their own sealed fuse just before the meter to protect their cable.

After the meter various circuits are formed in the junction box for different parts of the house and for different uses (figure 110).

Each circuit will have its own fuse of suitable value, 5 amps for lighting circuits, 13 or 15 amps for power circuits. Plugs for appliances often have fuses in them and these should be rated just above the maximum current the appliance could need. In this way the fuse in the plug will break without interrupting the rest of the circuit (figure 111).

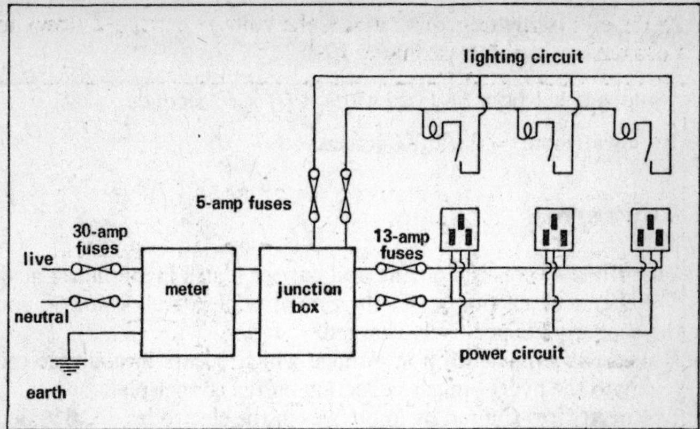

Figure 110. House wiring circuit

The earth wire

This is incorporated in the power circuits as a safety device. It is the green-and-yellow-striped wire and is connected to the outside metal casing of appliances. The appliance will always be at earth potential, the same potential as a person touching it. If a live wire should touch the case, a large current will pass to earth and the fuse will break.

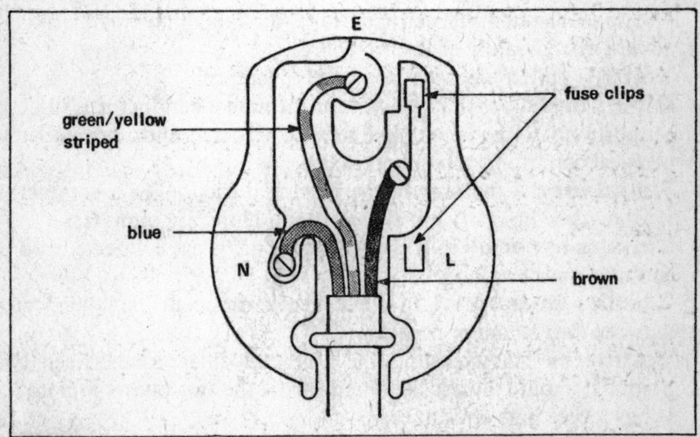

Figure 111. 13-amp plug

Kilowatt-hour

The joule is a rather small unit of energy. The kilowatt-hour is a larger, more convenient unit for electricity boards.

$$1 \text{ watt} = \frac{1 \text{ joule}}{1 \text{ second}}, \text{ thus } 1 \text{ joule} = 1 \text{ watt} \times 1 \text{ second}$$

$$1 \text{ kilowatt} \times 1 \text{ hour} = 1\,000 \text{ watts} \times 60 \times 60 \text{ seconds}$$

$$1 \text{ kilowatt-hour} = 3\,600\,000 \text{ joules.}$$

Key terms

Simple cell consists of zinc and copper plates in sulphuric acid. The chemical reaction leaves the zinc plate negatively charged, and the copper plate positively charged.
Polarisation The drop in voltage which occurs in a simple cell owing to the hydrogen gas collecting on the copper plate.
Local action Caused by impurities in the electrodes, local action results in slow eroding of the electrodes.
Daniell cell overcomes the erosion problem and polarisation.
Secondary cells (accumulators or **storage cells)** Cells in which the chemical change is reversible. Electrical energy can be put in, stored chemically, then taken out when needed.
Flow of charge through conductors constitutes an electric current. The conventional direction of current is shown by positive charge flowing from positive to negative. In fact we now know that it is negative electrons that flow from negative to positive.
Ampère The unit of electric current; the rate of flow of charge.
Coulomb The unit of electric charge.
Volt The unit of electrical potential difference.
Ohm's law states that the current through a conductor is directly proportional to the potential difference across its ends, provided the physical conditions remain constant.
Resistance A measure of the 'electrical friction' in a conductor. If Ohm's law holds then it is a constant. Units are ohms (Ω).
Circuits in series In these the components are connected end to end in a complete loop.
Circuits in parallel In these the components are connected between two common points.
Voltmeter This measures the potential difference between two points. It should always be joined across the two points in parallel. It has a very high resistance.
Ammeter This measures the current flowing through a wire. It should be placed in series with the rest of the circuit. It has a very low resistance.
Resistivity of a material is the resistance of a cubic metre of the material. Units are ohm metres (Ω.m).

Temperature coefficient of resistance The fractional increase in resistance compared with the resistance at 0°C for each degree change in temperature. Units are per degree centigrade (/°C).
Resistance thermometer consists of a coil of wire whose resistance at 0°C and temperature coefficient are known. Its resistance at any other temperature can then be used to calculate the value of that temperature.
Electromotive force (e.m.f.) of a cell The potential difference across the cell when there is no current being taken from the cell. Units are the same as potential difference, volts (V).
Internal resistance of a cell The resistance to the flow of charge in the cell itself.
Lost volts The term for the drop in potential difference across the terminals of a cell when current is taken from the cell. It is caused by the potential drop across the internal resistance.
Potentiometer A device for producing a continuous range of potential differences. With a uniform-resistance wire the potential difference is proportional to the length used.
Electrolysis The passing of electricity through liquids which produces a chemical reaction.
Electrolyte The liquid used in electrolysis.
Electrode The metal plate that conducts the electricity into the electrolyte.
Anode The positive electrode.
Cathode The negative electrode.
Ions The charge carriers that carry the current through the electrolyte. They may be positively charged, being atoms or molecules that have lost one or more electrons, or they may be negatively charged, being atoms or molecules that have gained one or more electrons.
Faraday's laws link the masses of substances involved in the chemical reactions of electrolysis with the total charge that has flowed and with their chemical equivalent weights.
Voltameter The vessel in which electrolysis takes place.
Corkscrew rule indicates the direction of the magnetic field around a straight wire carrying a current.
Solenoid A long thin coil. The shape of the magnetic field around one when it is carrying a current is similar to that of a bar magnet.
Electromagnet has a core of soft iron. The soft iron is magnetised by the small field produced by the current in the coil but loses its magnetism as soon as the current is switched off.
Motor effect The force on a wire carrying a current in a magnetic field.

Fleming's left-hand rule links the directions of the magnetic field, the current and the force on the conductor.
Galvanometer An instrument for detecting a small electric current.
Eddy currents The electric currents that are produced in a piece of metal when it is in a changing magnetic field.
Armature The rotating part of an electric motor.
Commutator The end of the armature that makes the electrical connections to the brushes.
Brushes The spring-loaded, carbon electrical contacts connecting with the rotating commutator.
Laminated soft iron core This is made up of thin sheets of iron, each electrically insulated to reduce the flow of eddy currents.
Electromagnetic induction The production of an electric current in a conductor in a changing magnetic field.
Faraday's law of electromagnetic induction links the e.m.f. produced with the rate of change of magnetic field.
Lenz's law indicates the direction of the induced current.
Fleming's right-hand rule links the directions of the induced current, the magnetic field and the movement of the wire.
Generator (dynamo) produces electrical energy by rotating a coil in a magnetic field.
Alternating current (A.C.) Current which flows first in one direction and then in the other. The mains supply in Britain alternates 50 times a second; it has a frequency of 50 Hz.
Direct current (D.C.) Current flows in one direction only. From a battery the current will have a steady value. From a D.C. generator its value tends to fluctuate.
Transformer A device for changing the voltage of an alternating supply. A step-up transformer increases the voltage, a step-down transformer decreases the voltage.
Induction coil A step-up transformer for producing very high-voltage sparks. It uses a switching D.C. supply.
Fuse The weakest wire in the circuit. If the circuit is overloaded the fuse will melt and break.
Kilowatt-hour The unit of energy used by electricity boards. It is the same as 3 600 000 joules.

Examples

Current electricity

(1) Find the resistance of three resistors, each of 2 ohms, connected in parallel.

$$\frac{1}{R} = \frac{1}{2} + \frac{1}{2} + \frac{1}{2} = \frac{3}{2}, \quad \text{so } R = \frac{2}{3} \text{ ohm}$$

The commonest error in this type of calculation is to express the answer as 3/2 ohms.

(2) If the resistivity of copper is $1{\cdot}8 \times 10^{-8}$ ohm.m, find the resistance of 10 m of copper wire of cross-sectional area 1 mm^2. Watch out for units in this type of question. Note that the cross-sectional area is given in mm^2, so it must first be converted to m^2.

$$1 \text{ mm} = 10^{-3} \text{ m}, \quad \text{so } 1 \text{ mm}^2 = 10^{-6} \text{ m}^2$$

$$R = \frac{\rho . l}{A} = \frac{1{\cdot}8 \times 10^{-8} \times 10}{10^{-6}} = 0{\cdot}18 \text{ ohm.}$$

(3) A meter has a resistance of 5 ohms and registers a full-scale deflection with 15 milliamps. How could the meter be adapted to read up to 1·5 amps?

The meter must be fitted with a shunt, placed in parallel with it. Only 15 milliamps (i.e. 0·015 amp) may pass through the meter. $\therefore$ 1·5 − 0·015 = 1·485 amps must be shunted around the meter.
The p.d. across the meter = current × resistance
= 0·015 × 5 volt
The p.d. across the shunt = current × resistance
= 1·485 × R
where R is the resistance of the shunt.

But these p.d.s must be equal, since meter and shunt are in parallel.

$$\therefore \quad 0{\cdot}015 \times 5 = 1{\cdot}485 \times R$$

$$\therefore \quad R = \frac{0{\cdot}015 \times 5}{1{\cdot}485} = 0{\cdot}0505 \text{ ohm}$$

Internal resistance

A battery charger has an e.m.f. of 14 volts and an internal resistance of 1 ohm. It is used to charge up a car battery of internal resistance 2 ohms and e.m.f. 12 volts. What is the charging current?

The charger must be placed in opposition to the battery, i.e., positive to positive. The net e.m.f. across the battery terminals is thus (14 − 12) = 2 volts, and the total resistance through which current must be driven is 3 ohms. The current is therefore 2/3 amp.

Heating effect

An electric cooker has an oven rated at 3 kilowatts, 4 plates rated at 1 kilowatt each and a grill rated at 2 kilowatts. The cooker works of 240-volt mains. Calculate:

(i) the total current which flows when all the parts are switched on;
(ii) the cost of using all the parts for three hours if electricity is charged at 2p per kilowatt hour;
(iii) the resistance of one of the plates.

(i) the total power consumption $= 3+4(1)+2 = 9\text{ kW}$
Since power = potential difference × current:
$9\,000 = 240 \times \text{current}$

$$\text{current} = \frac{9\,000}{240} = 37{\cdot}5 \text{ amps.}$$

(ii) Power consumption $= 9\text{ kW}$
$\therefore$ energy used = power × time $= 9 \times 3 = 27\text{ kWh}$
$\therefore$ cost at 2p per kWh $= 27 \times 2 = 54\text{p}$

(iii) Current through plate $= \dfrac{1\,000}{240}$ amps

$$\text{Resistance of plate} = \frac{\text{potential difference}}{\text{current}} = 240 \div \frac{1\,000}{240}$$

$$= \frac{240 \times 240}{1\,000} = 57{\cdot}6 \text{ ohms}$$

Chemical effect

What current is needed to deposit 4·02 g of copper in 2 hours if the e.c.e. of copper is $3{\cdot}3 \times 10^{-7}$ kg/coulomb?

$$\text{Current} = \frac{\text{mass deposited}}{\text{e.c.e. of Cu} \times \text{time in seconds}} = \frac{4{\cdot}02 \times 10^{-3}}{3{\cdot}3 \times 10^{-7} \times 2 \times 60 \times 60}$$

$$= 1{\cdot}7 \text{ amps.}$$

Note Examiners are particularly keen on questions involving two or more physics topics. Here is an example involving **energy changes.**

An electric motor takes a current of 2 amps at a potential difference of 24 volts. It is used to raise a load of mass 3 kg, which can be raised through a vertical distance of 6 m in 5 seconds.

(i) What is the power rating of the motor?
(ii) How much electrical energy is supplied to lift the load?
(iii) How much mechanical energy is used to lift the load?
(iv) What is the efficiency of the lifting system?
(Take $g = 10\ \text{m/s}^2$.)

(i) $P = IV = 2 \times 24 = 48$ watts
(ii) $E = IVt = 2 \times 24 \times 5 = 240$ joules
(iii) Mechanical energy = mass × acceleration due to gravity × height raised
$= 3 \times 10 \times 6 = 180$ joules
(iv) Efficiency $= \dfrac{\text{mechanical energy}}{\text{electrical energy}} \times 100 = \dfrac{180}{240} \times 100 = 75\%$

Chapter 9
Electronics

Electronics is the general term used to describe effects associated with the behaviour of **electrons** under certain circumstances. These include the passing of electrons through a vacuum, as in a **cathode-ray tube**, and the effects resulting from the flow of electrons through devices which do not obey Ohm's law, such as **valves** and **transistors**. One of the basic facts essential to the study of electronics is that the value of the charge carried by an electron is known. Historically this was first determined in an experiment carried out by Robert Millikan.

Millikan's experiment to determine the charge carried by an electron

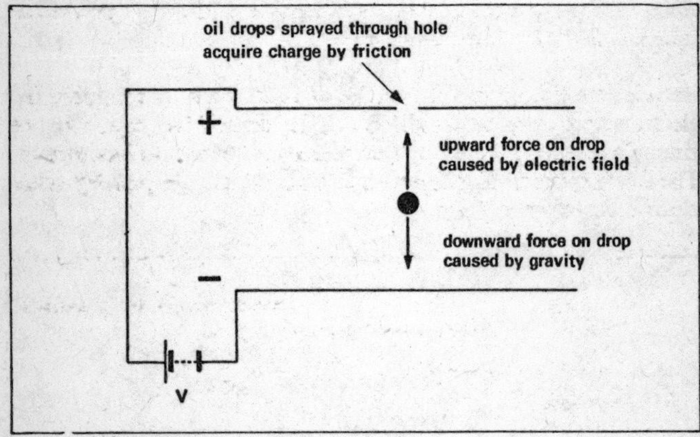

Figure 112. Millikan's oil drop experiment

This is the only experimental evidence that shows in a direct way that electric charge comes in 'packets'.

Millikan measured the charge carried by very small charged oil drops by suspending them in an electric field produced between two parallel plates (figure 112). Under these conditions, the upward force on the drop caused by the effect of the field is equal to the downward force on the drop caused by its weight.

He took many hundreds of readings and obtained a large number of values for the total charge carried by each of the drops. However, he found that all these values were simple multiples of one value, and that was **$-1 \cdot 6 \times 10^{-19}$ coulombs**.

Therefore he concluded that the smallest quantity of charge which could exist was $-1 \cdot 6 \times 10^{-19}$ coulombs, and that this was the charge carried by an electron.

Cathode-ray tube

When certain metals are heated, electrons are emitted from their surface. This process is known as **thermionic emission**. In the cathode-ray tube (figure 113) two electrodes are fixed inside an evacuated tube, one of which (the cathode) is heated and emits electrons that are attracted to the anode. The anode is at a very high positive potential so that the electrons are accelerated to a very high speed. The anode has a hole in it so that many of the electrons pass straight through producing a beam. This part of the tube is called the **electron gun**.

Between the anode and the screen at the far end of the tube, the electrons travel in a straight line. The screen has a coating of zinc sulphide and the electrons, on striking it, cause it to **fluoresce**. The electron beam may be deflected from its straight path by either electric or magnetic fields.

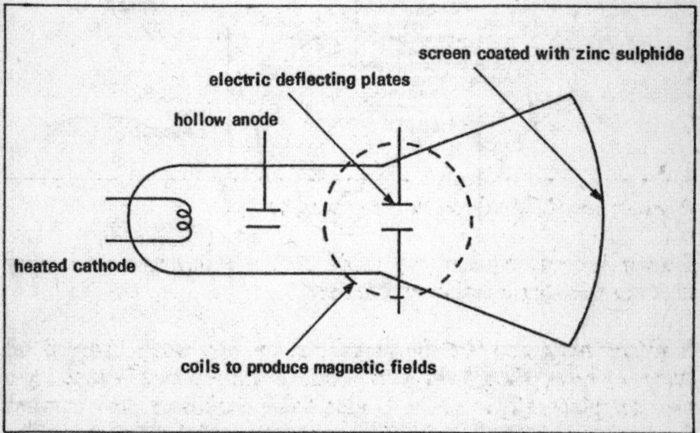

Figure 113. Cathode-ray tube

The electric field is produced by a potential difference across electrodes in the tube.

The magnetic field is produced by current flowing through coils situated outside the tube.

For the electrical deflection the electron beam is bent towards the positive plate.

The direction of the magnetic deflection is given by Fleming's left-hand rule, remembering that the conventional current flows in the opposite direction to the electron beam.

The thermionic diode

This consists of two electrodes in a glass 'envelope' from which all the air has been removed (*di* = two). One of the electrodes, the **cathode**, is heated by passing a small current through it (figure 114).

This heated cathode emits electrons. One can imagine them as being boiled off rather as steam is boiled off water.

If the other electrode, called the **anode**, is made to have a positive potential it will attract the electrons, and a current will flow.

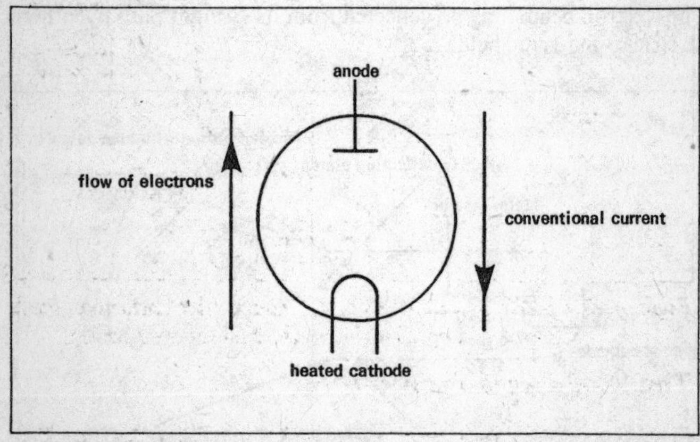

Figure 114. Thermionic diode

If, however, the anode is made negative, the electrons around the cathode will be repelled and no current will flow. Therefore the diode is a device that allows current to flow **in one direction only**: this is why it is sometimes called a **valve**.

Relationship between anode current and anode potential

This can be investigated using the circuit shown in figure 115a. A graph of anode current against anode potential is called the **characteristic** curve of the diode (figure 115b). The current reaches a maximum value called the **saturation current**. All the electrons emitted by the cathode are being collected by the anode. A higher temperature of the cathode will produce more electrons and thus a higher saturation current.

It might be expected that only a very small positive potential on the anode would attract all the electrons. This is not the case as the cloud of electrons round the cathode produces a **space charge** that tends to shield the cathode from the positive potential of the anode.

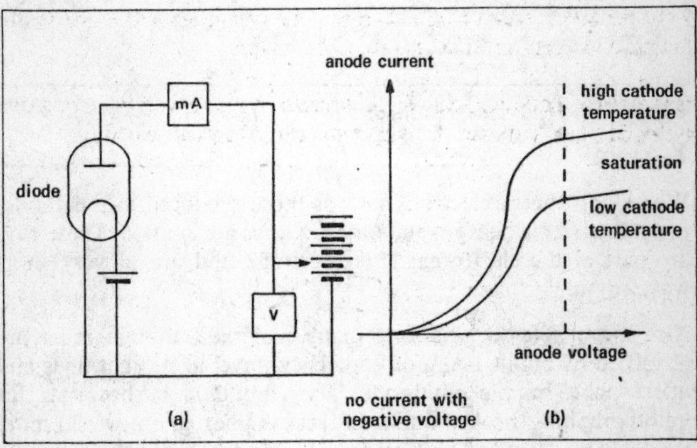

Figure 115. (a) Circuit to investigate relationship between anode current and anode voltage; (b) diode characteristics

The diode as a rectifier

One of the main uses of a diode is for rectifying A.C. to D.C. In the main, electrical power is produced with A.C. so that its voltage can be changed by transformers and transmitted at high voltages. However, there are many practical applications for D.C., such as radios.

The alternating voltage is applied across the diode, but current flows through it in one direction only. The undirectional voltage is taken from across a high resistance connected to the anode. Half-wave rectification is produced (figure 116). Two diodes working 'back to back' will produce full-wave rectification and by further smoothing true D.C. can be obtained.

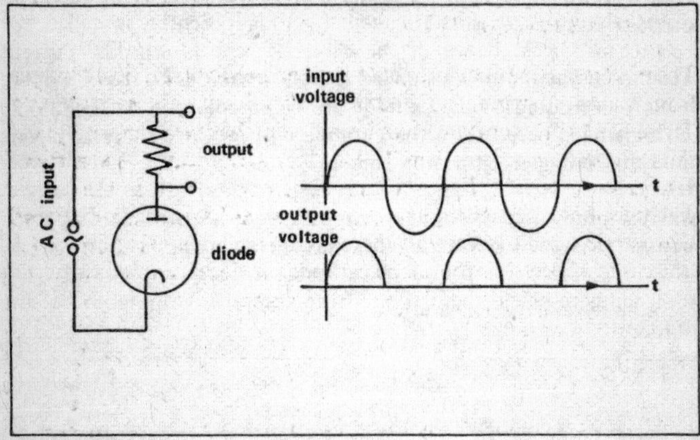

Figure 116. Half-wave rectification

X-rays

When high-energy electrons such as those produced in a cathode-ray tube strike a heavy metal target, X-rays are emitted. These rays are part of the electromagnetic spectrum, and are of very short wavelength.

X-rays are able to penetrate many materials, though most are absorbed by about 1 mm of lead. They travel in straight lines and affect photographic emulsions. They are used in hospitals for 'photographing' broken limbs. In fact the photograph would more precisely be described as a record of a shadow, since the rays pass through flesh, but not through bones.

Because the atomic spacing in crystals is about the same as X-ray wavelengths, they produce diffraction patterns when passed through the crystals, which act as diffraction gratings. From the resulting patterns the crystal structure can be worked out.

Each metal emits its own characteristic X-ray wavelengths. This information was used by Mosely in 1913 when sorting out the order of the elements in the periodic table.

Radio transmission

Radio waves are electromagnetic waves, similar to light but of much longer wavelength. The longest are hundreds of metres, the shortest of the order of a centimetre. Each broadcasting station transmits waves of its own particular wavelength. BBC Radio 2 broadcasts on 1 500 m wavelength, Radio 1 in the UK on 247 m. This is called the **carrier wave** (figure 117a).

The matter to be broadcast, whether speech or music, is first changed from sound vibrations to electrical vibrations by a microphone. These electrical vibrations will be at a much lower frequency and thus a much longer wavelength than the carrier wave. This **audio-frequency wave** (figure 117b) is superimposed on to the carrier wave. Figure 117c shows the resulting **amplitude-modulated wave**. It is this wave that is transmitted from the station aerial.

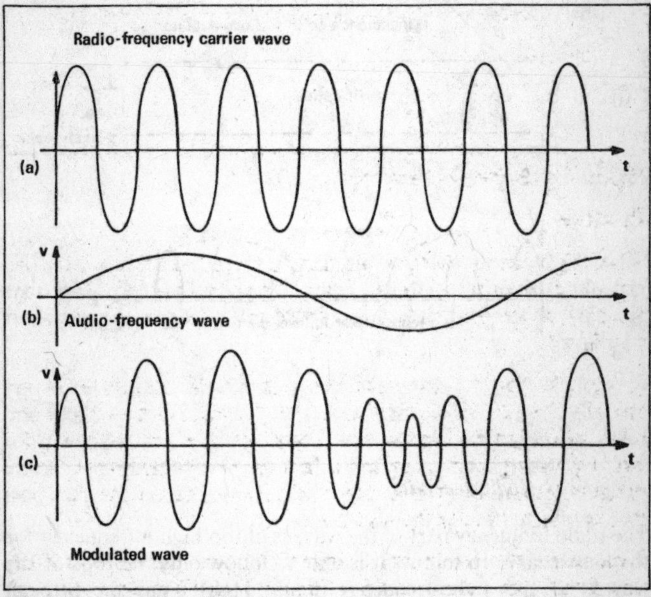

Figure 117. Radio transmission

Radio reception

The modulated radio wave is picked up by the aerial of the receiver and fed into a **tuned circuit**. But many other waves from other stations will also be picked up, all of different frequencies. The tuned circuit accepts the waves on one frequency and rejects the others.

The tuned circuit consists of an **inductance coil** and a **variable capacitor**. By varying the value of the capacitor it will **resonate** with different frequencies; this means that various stations can be tuned in. The wave is usually very weak and has to be **amplified**. The amplified wave passes through a detector, usually some form of diode, that allows the current to flow only in one direction, thereby cutting off the bottom half of the wave. It finally goes to the earphones or loudspeaker (figure 118a).

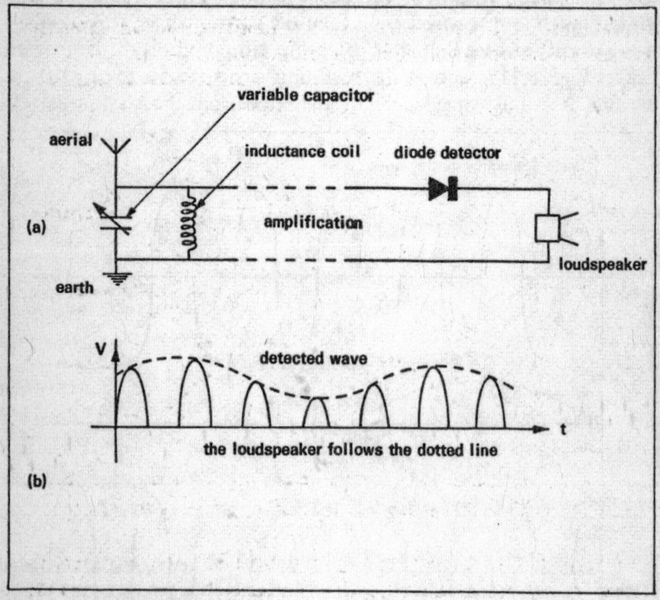

Figure 118. Radio reception

The radio frequency part of the wave is of too high a frequency for the loudspeaker to follow. It is able to follow only the tops of the waves, i.e. the audio-frequency (figure 118b). Thus the original sound is heard.

Transistors

These small electronic devices have revolutionised both the electronics industry and the everyday lives of millions. The transistor has made it possible for manufacturers to produce small, light amplifiers that do not waste energy in heating electrodes (as the old-type valve amplifiers used to). Transistors are also used in small, fast, switching circuits.

A transistor consists of three parts, the **emitter**, the **collector** and the **base**, to which leads are attached (figure 119a). A small flow of current into the base produces a large flow of current through the transistor from the emitter to the collector.

The resistance between the emitter and collector is high when there is no current flowing into the base (figure 119b). However, when a small current flows into the base the emitter-collector resistance decreases to a very small value (figure 119c).

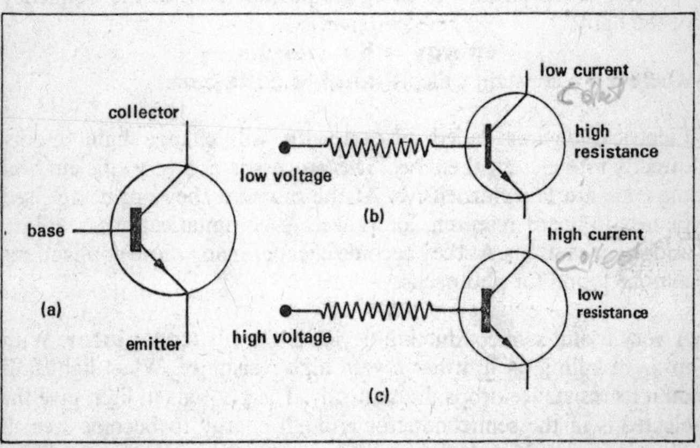

Figure 119. (a) Transistor symbol; (b) and (c) action of transistor

For amplification, the current on the base is varied around the critical value, thus producing large changes in the emitter-collector current. For switching, the base is either well above or well below the critical value. The switch is either on or off. This is the basis for all computer (binary) arithmetic, with each switch either 0 or 1.

Photoelectric effect, photocells and photoresistors

The **photoelectric effect** shows that in certain circumstances light has to be considered as packets of energy, **quanta** or **photons**, having a particle-like nature.

Light above a critical frequency falling on a metal surface will cause electrons to be emitted from the surface. The critical frequency varies from metal to metal: the more reactive the metal the lower the frequency.

The following simple experiment can be performed in school laboratories. A negatively charged electroscope connected to a clean zinc plate is rapidly discharged when the plate is irradiated with ultraviolet light. No such effect occurs if the electroscope is positively charged. The radiation releases electrons from the zinc atoms which are held back when the plate is positively charged, but are emitted when it is negatively charged.

The size of the packets of energy is proportional to the frequency of the light:

energy = h × frequency

where h is a constant called **Planck's constant**.

Electrical devices called **photocells** will change light energy directly into electrical energy. The process is not very efficient and the cells are fairly expensive. At the moment they are being used mainly in space research, for powering communications satellites and space-stations. As they become cheaper many more applications may be found for photocells.

A very useful semiconducting device is the **photoresistor**. With no light falling on it, it has a very high resistance. When light falls on it its resistance drops dramatically. The photons of light give the electrons in the semiconductor enough energy to become free. If used in a circuit with a transistor they make a very useful light-sensitive switch.

Key terms

Millikan's experiment measures the size of the charge carried by an electron.
Cathode ray The beam of electrons emitted by the cathode.
Electric gun The device that produces the cathode rays.
Thermionic diode 'Thermionic' indicates a heated cathode. 'Diode' means it has two electrodes.
Cathode The negative electrode.
Anode The positive electrode.
Saturation occurs when an increase in voltage produces no increase in current.

Characteristics of an electronic device are the set of current/potential difference graphs.
Space charge The cloud of electrons that 'boil' off the cathode when it is heated.
Rectifier An instrument which changes A.C. to D.C.
Diode allows electrons to flow in one direction but not the other.
X-rays The rays produced at the anode when hit by the cathode rays. They are electromagnetic waves of very short wavelength.
Carrier wave The radio wave transmitted by a radio station. Each station has its own wavelength and frequency.
Modulated wave The carrier wave which has the audio-frequency music or speech superimposed on it.
Tuned circuit The first part of the radio receiver that accepts the carrier wave of one particular wavelength and rejects the others.
Amplification increases the amplitude of the wave.
Detector cuts off one half of the carrier wave allowing the speaker to respond to the audio-frequency.
Transistor A semiconducting device with three electrodes used for amplification and switching.
Photoelectric effect The emission of electrons from metals when light falls on them. It shows that light energy comes in packets.
Quanta/photons are both names for the packets of energy associated with light.
Photocell A device that changes light energy to electrical energy.
Photoresistor The semiconducting device that has a high resistance in the dark, but a low resistance when light falls into it.

Chapter 10
Radioactivity

It has been found that certain of the heavy elements are unstable. They disintegrate spontaneously emitting particles and energy, changing into other elements in the process. Radioactive emissions are of three types:

α-particles

These are **helium nuclei**, helium atoms that have lost their two orbiting electrons. Therefore they are positively charged, consisting of two protons and two neutrons (see page 128). Though they are ejected from the nucleus of the radioactive substance with considerable energy, they are relatively large particles and are easily absorbed by matter, e.g. a few mm of air, cardboard or tinfoil.

A nucleus that ejects an α-particle will form a new nucleus with a mass number reduced by 4 and atomic number reduced by 2. E.g. radium ejects an α-particle and becomes an inert gas, radon, as follows: ${}^{226}_{88}Ra \rightarrow {}^{222}_{86}Rn + {}^{4}_{2}He$. In this equation the subscript denotes the atomic number of the nucleus, while the superscript denotes its mass number.

β-particles

These are **electrons**, not from the orbiting electrons of the atom but from within the nucleus. They are ejected with great speed and have a range of up to 15 cm in air.

A nucleus that ejects a β-particle will form a new particle with the same mass number but with one added to its atomic number, e.g. thorium ejects a β-particle and becomes protactinium:

${}^{234}_{90}Th \rightarrow {}^{234}_{91}Pa + {}^{0}_{-1}e$.

γ-rays

These are very high energy **electromagnetic waves** that are emitted at the same time as the α and β particles. They are similar in nature to light but of very much shorter wavelength. They are very highly penetrating, being capable of passing through several centimetres of lead.

Half-life This is the time taken by half of a sample of a radioactive element to decay into a new element. This new element may or may not be radioactive. Half-lives vary from $4{\cdot}6 \times 10^9$ years for uranium 238 to fractions of a second for others.

Methods of detecting radioactive emissions

Photographic emulsions

The particles affect the emulsion in a similar way to light. The radioactive substance may be outside the light-tight wrapper but will still affect it, as the particles and γ-rays will pass through the wrapper. Historically this was the first evidence of the existence of radioactivity.

The other methods depend upon the fact that the α and β-particles, when passing through the air, knock some of the electrons off the air molecules, leaving a trail of positive and negative charges, or **ions**, causing the air to become conducting.

Gold-leaf and pulse electroscopes

It was seen on page 129 that when the gold leaf is charged it diverges and as the central stem is insulated it will remain diverged till the leaf is somehow discharged. If a radioactive source is brought close to the instrument, the leaf will slowly fall. The air will have become a conductor and the charge on the leaf will leak away.

The pulse electroscope works in the same way, but as soon as it discharges it automatically charges itself again. The number of times it does this is a measure of the amount of ionisation and thus a measure of the amount of radioactivity present.

Spark counter and Geiger-Müller tube

These instruments indicate the presence of individual particles.

The **spark counter** consists of a thin wire a few millimetres away from a flat plate. The wire is at a high potential relative to the plate and almost on the point of sparking. When a single particle passes between the wire and the plate it reduces the insulation of the air sufficiently for a spark to cross.

The **Geiger-Müller tube** is similar to the spark counter, but increases the volume that is sensitive to the presence of ionising particles. A central rod that is kept at a few hundred volts is surrounded by a conducting cylinder.

The tube is connected to an electronic counter and the arrangement can often count thousands of particles a second (figure 120a).

Cloud chamber

This instrument not only detects individual particles but also provides a record of their tracks.

By sudden expansion alcohol or water vapour in the chamber is cooled until it is **'supersaturated'** (see page 74). Some of the vapour should change back into the liquid state, but for the vapour to form drops, it needs dust particles to 'start the drops off'. If there is no dust present the drops will form on ionised particles instead.

If α or β particles pass through the chamber when it is in this supersaturated state they leave a trail of ionised particles behind them on which the drops will form. These drops are big enough to be seen and photographed.

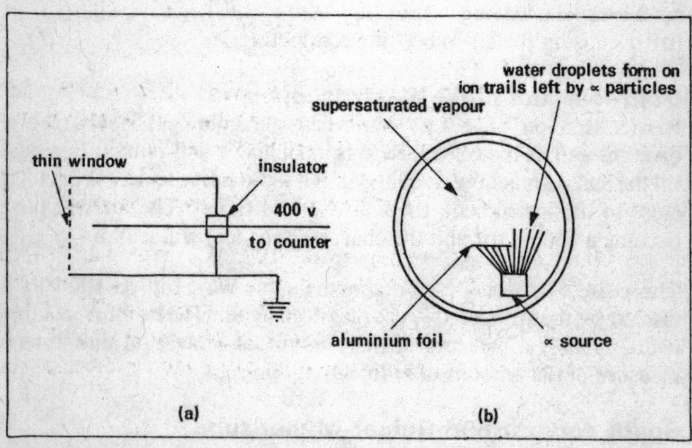

Figure 120. (a) Geiger-Müller tube; (b) cloud chamber

Figure 120b shows α-particle tracks in a cloud chamber showing their range in air and that they are stopped by a thin piece of aluminium foil.

Radioactive isotopes

Many elements are made up of two or more isotopes. **Isotopes are atoms with the same atomic number but different mass number.** They have the same number of protons in the nucleus and therefore the same chemical properties. But they have a different number of neutrons in the nucleus. The number of neutrons in a nucleus affects its stability, and some of the heavier ones are unstable and thus radioactive. Radioactive isotopes of many of the lighter elements have been produced artificially. These have many uses in research, medicine and industry.

Key terms

α-particles Helium nuclei: atomic number 2, mass number 4.
β-particles Electrons: atomic number −1, mass number 0.
γ-rays Very high energy electromagnetic waves.
Half-life The time taken for half the mass of a radioactive element to decay.
Isotopes Atoms with the same atomic number but different mass numbers. They have the same chemical properties.

Example

Radioactive decay

(i) Radium nuclei, of mass number 226 and atomic number 88, decay by the emission of an alpha particle into radon. What are the mass number and atomic number of the radon nuclei?
(ii) In what way does the nucleus of a radioactive atom change as the result of emitting an α-particle followed by two β-particles?

(i) An alpha particle consists of 2 protons and 2 neutrons, so has an atomic number of 2 and a mass number of 4. Emission of an alpha particle will therefore reduce the atomic number by 2 and the mass number by 4. So radon must have a mass number of 222 and an atomic number of 86. This type of reaction may be represented in equation form:

$${}^{226}_{88}Ra \rightarrow {}^{222}_{86}Rn + {}^{4}_{2}He$$

The superscript (top number) represents the mass number and the subscript (bottom number) represents the atomic number.
(ii) a beta particle is an electron. It has a negligible mass and a charge of −1. Representing a β-particle by ${}_{-1}^{0}e$:

$${}^{A}_{Z}X \rightarrow {}^{4}_{2}He + 2\,{}_{-1}^{0}e + {}^{A-4}_{Z}Y$$

The emission of beta particles makes no difference to the mass number (^{A}), but increases the atomic number by 1 for each β-particle emitted.
Thus X and Y have the same atomic number, but different mass numbers. They are therefore different isotopes of the same element.

Index

Examination Hints

Answering examination questions

Physics is a precise subject. It is concerned with precise relationships and uses precise language.

Archimedes' principle, for example, may be stated: 'When a body is wholly or partly immersed in a fluid, the upthrust on the body is equal to the weight of fluid displaced.' Often in examination papers one comes across such answers as: 'The upthrust on a body equals the amount of water displaced.' Notice how much better is the first definition. Firstly it uses the precise word 'weight' instead of the vague word 'amount', which could be taken to mean mass, weight or volume. Secondly, it refers to all the cases in which the principle may be applied—partly or wholly immersed, and in all fluids. The second definition suggests that the principle applies only to water. While the second definition is not actually wrong, it would score only about one-quarter of the marks available in an examination question, because of its vagueness. In general, avoid using words like 'quantity' and 'amount', which are imprecise.

If asked, for example, to state the relationship between the length of a stretched wire and the frequency of the note it produces when plucked, it would be true to say that the frequency depends on the length, but again this is an imprecise statement. For full marks, the relationship should be stated precisely, i.e. the frequency is inversely proportional to the length. Similarly, the electrical resistance of a wire is directly proportional to its length and inversely proportional to its cross-sectional area.

Some common words are frequently confused by candidates. For example, the words 'sensitive' and 'accurate' are often wrongly used. A pair of household scales can be an **accurate** weighing device, but they are not very **sensitive**. They certainly could not be used to weigh a pin. For this, a balance capable of detecting and measuring a small weight is required. The word 'sensitive' applies therefore to things which will detect small changes. A clinical thermometer is a sensitive instrument because it will detect small changes in temperature—changes of about 0·1°C. A domestic thermometer, such as might be used to measure room temperature, will not do this. A milliammeter will detect changes in current of 0·001 amp, but the sort of meter found in motor cars is much less sensitive, and would not respond to such a small change. Nevertheless, the room thermometer and the car ammeter may both be accurate instruments. Conversely, a milliammeter may not be particularly accurate.

The word 'induction' is also a frequent cause of difficulty. Its use in the contexts of magnetism and electricity means 'without touching' or 'without contact'. If asked to describe the charging of a gold-leaf electroscope by induction, you should describe in your answer how the charged rod is brought close to, but not touching, the electroscope. Candidates frequently make the mistake of saying, 'Touch the electroscope with the charged rod', which immediately loses them all the marks for that section of the question.

Attention to detail

Many questions in physics, in both course work and examinations, ask for written descriptions of how you would carry out an experiment. When your teacher, or the examiner, marks your answer, he will probably consider four points, and will look to see how well you have covered each of them.

(1) The apparatus There is no need to make a list of the apparatus used, although some candidates do, and it is quite a good idea in that it helps you to think straight about the rest of your answer. Similarly, there is no need to describe simple apparatus in very much detail. But any complicated and special apparatus—such as that used for Boyle's law—should be described and a diagram drawn.

(2) The method This should be a description of what is done—how the apparatus is used and what measurements are made. It is here that marks are often lost for inattention to detail. In some cases, for example, the examiner will want to know not only what was done but how it was done. It is not sufficient to say, 'A metal block was heated and dropped into cold water', without mentioning how the metal block was heated. In some cases, of course, it may be clear from your diagram, but it is still important to add, in the above case, that the block was heated for sufficient time to allow the whole of the block to reach a steady temperature.

(3) Results Assuming that your method has included details of what measurements to make and how to make them the next task will be to explain what needs to be done in order to work out the result. In an experiment to investigate Boyle's law, for example, after a series of values of pressure and volume has been obtained, a graph of p against $1/V$ should be plotted and it should be found to be a straight line. Alternatively, it could be stated that the pressures should be multiplied by their corresponding volumes, in which case the products would be expected to be constant.

(4) Precautions In some questions you may be asked under a separate heading to list any precautions which should be taken in order to make results more accurate. Even if this is not the case, there will often be a mark or two reserved for them and they should be mentioned, either at the end or in the body of the answer. An obvious point, common to a large number of measurements, is that the observations should be repeated and an average taken of the results. In heat experiments, stirring and lagging are the two commonest precautions to mention.

Diagrams

Diagrams should always be used where they will help make answers clear. It is a good idea to practise drawing diagrams, both in order to improve your speed, and also as a good method of revising a particular topic. Always make sure your diagrams are large, clear and labelled. When drawing diagrams of electrical or mechanical devices, such as pulley systems, make sure that they will work. Imagine yourself pulling on the string and try to assess whether or not the load would be raised. If not, your diagram is obviously not a good one.

Some topics call for **scale diagrams**. You are most likely to meet these in questions dealing with forces—especially those in which the resultant of two forces has to be found—and in problems involving lenses or mirrors. Apart from making sure you know how to tackle this type of question, always make your diagrams large, state the scale, and use a hard sharp pencil to improve the accuracy. It is often a good idea, in optics diagrams, to use a larger scale for the height of the object than for its distance from the lens or mirror.

Note on mathematics

Physics often makes use of mathematics in order to express relationships more precisely, and this is reflected in the number of calculations which are normally found in both course work and examination papers. A full revision of all the mathematical topics which you might need to know is beyond the scope of this book, but apart from purely arithmetical errors, there are two errors which seem particularly common in physics papers.

The first involves the use of algebraic formulae. A particular case is the formula for electrical resistors in parallel, i.e.

$$\frac{1}{R}=\frac{1}{R_1}+\frac{1}{R_2}+\frac{1}{R_3}$$

Perhaps the reason why so many candidates seem to make mistakes with this formula is that they try to take short cuts with it. A worked example, showing you how to find R if you are given values for R_1, R_2 and R_3, will be found on pages 162–3.

Many candidates have difficulty in handling very large or very small numbers, which are often written in standard form. Thus:
$3\,600\,000 = 3{\cdot}6 \times 10^6$ and
$0{\cdot}000\,000\,2 = 2 \times 10^{-7}$
It is important that you should be able to handle numbers of this type, and it is when such numbers are divided that most of the trouble seems to occur. If the potential difference between the plates of a capacitor is 3 000 volts, and the charge is 6×10^{-3} coulomb, the capacitance will be

$$\frac{6\times10^{-3}}{3000} = 2 \times 10^{-6}\text{ F.}$$

Arithmetical errors normally lose one mark in a calculation on an examination paper. Unfortunately, an arithmetical slip early on in a question can lead to difficulties at a later stage, so you are urged to pay particular attention to your arithmetic, however simple it may appear. Logarithms or slide rules may be used in examinations, but quite often the calculations set are designed so that the numbers are easy and the use of tables unnecessary.

Calculations

Calculations form a very important part of physics syllabuses, and candidates for examinations should feel confident of being able to handle them. One of the main reasons for this is that they enable marks to be gained relatively easily.

Many candidates find these calculations difficult, though hopefully after working through this book you will feel more at home with them. Apart from arithmetical errors, mistakes in calculations seem to arise in examinations for one or other of the following reasons:

(1) Inability to remember or use the appropriate formulae

It is surprising how many candidates in examinations fail to gain marks for calculations simply because they do not know the work properly. The equations of uniformly accelerated motion, for example, are often quoted incorrectly, resulting in no marks at all for the calculation in which the equations occur. In using the general gas equation:

$\frac{P_1 V_1}{t_1} = \frac{P_2 V_2}{t_2}$ it must be remembered that the temperatures must be measured in degrees Kelvin and not in degrees Celsius. An error of this sort would show the examiner that the basic principle behind the gas laws has not been understood, and such an error would be very heavily penalised. Only you can learn these points and we hope that this book will help you to revise the most important of them.

(2) Leaving out parts of a calculation

A particular example occurs with calculations involving method of mixtures, where it is quite easy to miss out one of the terms in the equation. You would lose about half marks in a question for this type of error.

(3) Failure to think out and express working fully and logically

On some candidates' scripts, answers to calculations appear out of thin air, with little or no working shown. Some examining boards state at the head of their papers that full marks for a calculation may only be gained if full working is shown, but even if this instruction is not specifically given, marks are likely to be deducted for work which is not fully explained.

Units

The answer to a numerical problem is meaningless unless the correct units are also stated. Now that most examining boards have adopted S.I. units, the problem is simpler, but the appropriate units for the various quantities likely to occur still need to be learnt. Some of the common ones are given below, with the abbreviation in brackets.

Mass	kilogram (kg)	Electric current	ampère (A)
Length	metre (m)	Potential difference	volt (V)
Time	second (s)	Resistance	ohm (Ω)
Force	newton (N)*	Charge	coulomb (C)
Work	joule (J)	Capacitance	farad (F)
Energy	joule (J)	Volume	cubic metre (m^3)
Power	watt (W)		

* Note that weight is a special kind of force, so weight will also be measured in newtons.

Make sure that you are familiar with the prefixes which are also used in S.I. The wavelength of light, for example, is normally given in **nanometres**, there being 10^{-9} nm in 1 metre.

It is not really necessary to learn a list of derived units, since these can be built up from those given above. Momentum, for example, is defined as mass × velocity, so its units will be those of mass multiplied by those of velocity—i.e. kg.m/s. Note the use of the . to signify multiplication, and of the / to signify division. The moment of a force about a point is found by multiplying the force by the perpendicular distance between the point and the line of action of the force, so the units of a moment will be newton.metres (N.m). Many candidates state this unit incorrectly as N/m.

Descriptive questions

Another type of question which occurs quite often on physics papers is the descriptive question. Although no calculations are required, considerable care is nonetheless required for a good mark to be obtained, as the following example should serve to demonstrate.

Principles of heat transmission: explain the parts played by conduction, convection and radiation in the water-cooling system of a motor car.

Notice that what is asked in this question is the application of the scientific principles of conduction, convection and radiation, and not just simply a description of the cooling system. Most of the marks would be given for reference to the terms in the various parts of the system. In the following answer, it is the correct use of the terms in bold type which would score most of the marks.

Heat is produced in the cylinders and **conducted** through the walls to the water which surrounds them. The heated water rises by **convection** and passes through the top hose to the top of the radiator. Here the heat is **conducted** through the metal wall of the radiator and lost to the air by **conduction** and **convection**. Very little radiation occurs unless the temperature of the radiating body is very high. As the water cools it falls to the bottom of the radiator and is recirculated to the cylinder block. A water pump aids circulation, and a fan is used to pass air around the radiator.

Lastly, remember no one can sit the examination for you; your success will largely depend upon the time and effort that you are prepared to put into your work. Learn the facts of the subject thoroughly, think about them, try to understand them and be prepared to devote many joules to your effort.

Key Facts Revision Section

Properties of matter

Evidence of the atomic theory of matter comes from the crystal structure of many solids. The crystal shapes can be built from linking layers of spheres.

Liquids have a fixed volume but no fixed shape, therefore the atoms or molecules in a liquid attract each other but are able to move round each other.

Gases have no fixed shape nor volume. The molecules do not attract each other but can move freely.

Kinetic theory of heat states that heat produces **random motion of the atoms or molecules**. In solids the atoms or molecules vibrate and as the temperature rises they vibrate more violently. This explains why solids expand when heated.

When a solid changes into a liquid the crystal bonds are broken. This accounts for the latent heat energy of fusion.

In liquids and gases the atoms or molecules are in **continual random motion**, which increases in accordance with increase in temperature. This explains diffusion, the mixing of two gases or two liquids, and thermal expansion. A gas produces a pressure on the walls of its container because its molecules set up a continual random bombardment on the walls.

Cooling in a liquid is caused by evaporation from its surface. According to kinetic theory, the high-energy, fast-moving atoms or molecules escape the surface tension forces and become gas. The lower-energy atoms or molecules remain in the liquid, therefore the average energy decreases. But it is the average energy that determines the temperature.

Increased temperature in a gas causes the molecules to move faster. This results in either increase in pressure, or increase in volume, or both.

Brownian motion Direct evidence that atoms or molecules are constantly moving can be obtained by observing under a microscope small particles suspended in a liquid or gas. The small particles will be affected by the random bombardment of the atoms of the fluid and can be seen to move.

The oil film experiment indicates the size of molecules. The oil molecule is about 2×10^{-9} m in diameter.

Hooke's law states that provided the elastic limit is not exceeded the deformation of a material is proportional to the force applied to it. Therefore a graph of load against extension will result in a straight line through the origin.

Pressure $= \dfrac{\textbf{force}}{\textbf{area}}$ **Units: newtons/metre²** (N/m²).

In liquids:

pressure = depth × density × acceleration caused by gravity.

A mercury barometer measures atmospheric pressure in terms of the height of a column of mercury which the atmosphere can support. The space above the mercury is empty (a vacuum) and therefore exerts no pressure.

U-tube manometers measure differences of pressure existing on the liquid surfaces in either arm.

Boyle's law states that the volume of a fixed mass of gas is inversely proportional to its pressure, provided the temperature remains constant.

A graph of pressure against volume produces a curve that does not cross either axis.

A graph of pressure against 1/volume gives a straight line through the origin, thus: **pressure × volume = constant**.

Density $= \dfrac{\textbf{mass}}{\textbf{volume}}$ **Units: kilograms/metre³** (kg/m³).

Archimedes' principle states that when a body is wholly or partially immersed in a fluid it experiences an upthrust equal to the weight of the fluid displaced.

If a body is floating, then the upthrust is equal to the weight of the body.

Because 1 gram of water has a volume of 1 cm³ (1 000 kg has a volume of 1 m³), the volume of an object can be found by weighing in air and then in water. The upthrust in grams is numerically the same as its volume in cm³.

Motion

Velocity and acceleration

Velocity is defined as **speed in a definite direction**, or **displacement/time**. Both displacement and velocity are **vector** quantities: that is, they need both a magnitude and a direction in order to be fully defined.

A body moves with **uniform velocity** if it has a **constant rate of change of displacement**. For velocity to be uniform, note that both speed and direction must remain constant. Velocity, like speed, will normally be measured in metres/second (m/s).

Acceleration is defined as the **rate of change of velocity**, i.e. the change in velocity/time for change to occur. **Uniform acceleration** is defined as a **constant rate of change of velocity**. Acceleration is also a vector, and is normally measured in metres/second^2 (m/s^2).

The **acceleration caused by gravity** (g) is the acceleration of a freely-falling body. It does not depend on the weight of the body and has a value of about 9·8 m/s^2. For calculation purposes g is often taken as 10 m/s^2.

The equations of uniformly accelerated motion are as follows:

(1) $v = u + at$	u = initial velocity
(2) $s = \frac{1}{2}t(u+v)$	v = final velocity
(3) $v^2 = u^2 + 2as$	a = acceleration
(4) $s = ut + \frac{1}{2}at^2$	t = time (seconds)
	s = distance (metres)

If a body is decelerating, a will be negative. Therefore a body thrown vertically upwards may be said to decelerate at 9·8 m/s^2, and for calculations it can be taken that $a = -9{\cdot}8$ m/s^2.

Velocity/time graphs

Whether specifically requested in a question or not, it is often useful to draw a velocity/time graph. Always plot velocity along the vertical axis and time along the horizontal axis. The **acceleration** may be found by calculating the **gradient of** the graph, and the **distance** travelled by calculating the **area under the graph**.

Force, mass and acceleration

Examination candidates are often confused about these closely-related topics, therefore it is recommended that chapter 2 be worked through at least twice since some of the principles mentioned later in the section may make the earlier ones easier to understand.

Forces

Forces may be thought of simply as pushes or pulls. They may not be mechanical in origin (magnetic and electrical forces are examples of non-mechanical forces). Forces are **vector** quantities, needing both magnitude and direction to be defined fully. Forces may be represented by straight lines, the length of the line indicating the magnitude of the force and the direction of the line, shown with an arrow, indicating the direction of the force.

Weight is a special example of a force, being defined as the **strength of the earth's pull on an object**. The weight of an object can be considered to act at a single point, called the **centre of gravity**. The weight of the body varies according to where it is. Far out in space the body is **weightless**.

If a body is at rest, or moving with uniform velocity, then it will continue to do so unless a resultant force acts on it. This is **Newton's first law of motion**. The term 'resultant' means, in this case, **unbalanced**. If a resultant force acts on a body, the velocity of the body will change.

Forces are measured in **newtons** (N). Notice that this means that the **weight of a body should also be expressed in newtons**.

Inertia and mass

We have just said that bodies moving with uniform velocity resist changes which we may try to make to their motion in the sense that we have to apply a force to bring about these changes. This resistance to change in velocity is called **inertia**. Safety-belts in motor cars serve to protect passengers from the effects of inertia. If the car is braked hard, the passengers resist the attempt to change their velocities and in the absence of safety-belts they could be thrown sharply against, if not through, the windscreen. Because they are **inert**, the passengers tend to continue to move forward, and in order to stop them doing this the safety-belt is used to exert a restraining force on them.

A heavy roller is more difficult to push than a light one because the former has more inertia. This might lead you to think that inertia is a property associated with weight. But a spacecraft, far out in space, is weightless, yet still needs to have forces exerted on it in order to change its motion. It has no weight, yet it still has inertia. The property of the spacecraft—or any other body—which determines its inertia is called its **mass**.

Mass may crudely be defined as the quantity of stuff in a body. It is a **scalar** quantity and is measured in **kilograms**. Unlike its weight, the mass of a body does not change from place to place.

The equation $F = ma$ relates the force applied to a body, F, to the mass, m, and resulting acceleration of the body, a. This follows from **Newton's second law of motion**.

Examination questions often require the candidate to associate this equation with the equations of motion discussed in the previous section.

Balances

A **spring balance measures weight**, because it requires the application of a force to stretch the spring. Nevertheless, some spring balances are calibrated in mass units. Such balances should only be used at the place where they are calibrated, although the error introduced by small movements on the earth's surface is small.

A **beam balance compares masses**, because any variations in the gravitational pull would affect both sides of the balance equally. Masses may also be compared with an inertia balance.

Composition and resolution of forces

The **resultant** of a system of forces is the single force which has, in magnitude and direction, the same effect as the system of forces.

The **parallelogram of forces** rule states that if two forces acting at a point are represented in magnitude and direction by the adjacent sides of a parallelogram, the resultant may be represented by the diagonal of the parallelogram drawn from the point of contact of the forces.

A single force may be **resolved** into two **components** at right angles to each other. The magnitude of the component is equal to the product of the force and the cosine of the angle between the force and the component.

Newton's laws of motion

(1) Every body continues in its state of rest, or uniform motion in a straight line, unless there is an unbalanced external force acting on it.

(2) The rate of change of momentum of a body is proportional to the unbalanced force acting on it and takes place in the direction of the force.

(3) To every action there is an equal and opposite reaction.

Momentum and impulse

The momentum of a body is defined as the **product of its mass and velocity**. Momentum, because it involves velocity, is a vector quantity. The units of momentum are those of mass multiplied by those of velocity, i.e. **kg.m/s**.

The first equation of motion, $v = u + at$, and the force equation $F = ma$, may be rewritten as

$$a = \frac{v-u}{t} \quad \text{and} \quad a = \frac{F}{m}. \quad \text{Hence} \quad \frac{v-u}{t} = \frac{F}{m}$$

or $\quad Ft = mv - mu.$

The product Ft is called the **impulse of the force**.

The right-hand side of the equation is equal to the **change in momentum** which takes place when the velocity of a body is increased from u m/s to v m/s.

Writing the equation as $F = (mv - mu)/t$ is a mathematical way of expressing **Newton's second law of motion**.

Conservation of momentum

The law of conservation of momentum states that, in the absence of external forces, **momentum in a particular direction is conserved**. Questions set on the use of the law usually involve explosions or collisions.

Work, energy and power

Work is done when a force moves its point of application, and is defined as the **product of the force and the distance moved in the direction of the force**.

Work is a **scalar** quantity. The units of work are those of force (newtons) multiplied by those of distance (metres), i.e. newton.metres. These units are given a special name, **joules**, abbreviated to J.

Suppose a body of mass m kg is at rest on the ground. The weight of the body is mg newtons, where g is the acceleration caused by gravity. To lift the body we have to apply an upward force in order to counteract its weight, which is acting downwards. If we lift the body at a steady speed, this upward force will also be equal to mg newtons, since, by Newton's first law, there will be no resultant force acting.

Suppose the body is lifted through a vertical height h metres.

The work done = force × distance = mgh joules.

Lifting has given the body **energy**. Energy means the ability to do work, and if the body were, for example, a bucket of water, the water could do work in turning a paddle wheel if it were allowed to fall again.

This type of energy is called positional or **potential energy**. Its value is **mgh joules**. Energy, like work, is a **scalar** quantity.

Suppose the body is released and falls back to earth. As it falls it will lose its potential energy which is converted into motion, or **kinetic energy**. Just before it hits the ground it has lost all its potential energy, and its kinetic energy has now reached its maximum value. The velocity v with which it will hit the ground after falling from a height h can be calculated by using the equation of motion $v^2 = u^2 + 2as$. In this case, $u = 0$, $a = g$, and $s = h$.
So $v^2 = 2gh$
If both sides of the equation are multiplied by m, and divided by 2, the result is
$\frac{1}{2}mv^2 = mgh$

A very important law in physics is the **law of conservation of energy**. This states, briefly, that **energy can neither be created nor destroyed**. So if mgh joules of potential energy have been used up, they have been converted into $\frac{1}{2}mv^2$ joules of kinetic energy.

Note that kinetic energy is also a **scalar** quantity. Candidates often go wrong on this point, probably because they assume that since velocity is involved, kinetic energy must be a vector. But although velocities can be either $+v$ or $-v$, v^2 will always be positive, and that is why kinetic energy is a scalar.

Potential and kinetic energy are examples of **mechanical energy**. Energy can however exist in many different forms, such as **heat**, **light**, **sound**, **electrical**, **nuclear**, etc. There are many ways in which energy can be transformed from one form into another. In all cases the law of conservation of energy must be obeyed.

Power

Power is defined as the **rate of doing work**, or by the ratio **work/time**. Power is also a **scalar** quantity. The unit of power is the joule/second, and this too is given a special name, the **watt** (W).

Motion in a circle

The **centripetal force** produces the **acceleration towards the centre of the circle.** If that force is removed then the particle continues to move in a straight line, i.e. **along the tangent to the circle**.

An astronaut in a space laboratory that is circling the earth has the impression of weightlessness, since both the man and the laboratory experience the same accelerating forces.

Heat

Thermometers

The **Celsius** or **centigrade** thermometer has as its lower fixed point the temperature of pure melting ice, which is given the value 0, and as its upper fixed point the temperature of steam above water that is boiling at standard atmospheric pressure, which is given the value 100.

The **Kelvin** scale measures absolute temperature. The absolute zero, when all possible energy has been removed from a perfect gas, is 0 on this scale, or -273°C.

0°C is 273 K, 100°C is 373 K. **K** stands for **degrees Kelvin**.

The **clinical thermometer** has a number of special features:
(1) it has a limited range, from 35°C to 43°C;
(2) it has a fine capillary giving readings accurate to 0·1°C;
(3) a constriction allows the maximum value to be read, even after the thermometer has been removed from the patient.

Six's maximum and minimum thermometer uses alcohol as the expanding fluid while the length of mercury in the tube moves iron indicators which give the maximum and minimum readings.

Expansion

Expansion in solids, liquids and gases on heating is caused by the increased energy of the molecules, which move about more vigorously and as a result take up more room.

Coefficient of linear expansion is the fractional change in length per degree rise in temperature. **Units are /°C or /K.**

$$\alpha = \frac{\textbf{increase in length}}{\textbf{original length}} \times \frac{\textbf{1}}{\textbf{temperature change}}$$

Change in density: convection currents

If the volume of a substance increases, its density decreases. A less dense substance floats on one which is more dense. Thus hotter water rises to the surface. Hot air also rises and cold air falls to take its place.

The unusual expansion of water

As water is heated from 0°C to 4°C it contracts and therefore becomes more dense. Water is at its most dense at 4°C. Therefore convection currents travel in the reverse direction from 0°C to 4°C, the cold water coming to the surface and the warmer water sinking to the bottom.

Expansion of a gas

When a gas is heated the molecules move faster and usually the pressure and the volume will increase. Charles kept the pressure constant and measured the increase in volume as the temperature increased. He found that all gases had the same coefficient of expansion.

Charles' law

All gases increase by 1/273 of their volume at 0°C for each degree rise in temperature, provided the pressure is kept constant.

Extending Charles' law to temperatures below 0°C, a gas will decrease by 1/273 of its volume at 0°C for each degree drop below 0°C. Therefore by −273°C it will have zero volume.

Using the Kelvin scale of temperature, Charles' law can be restated more simply: at constant pressure, the volume of a gas is proportional to its Kelvin temperature;

or: $$\frac{V_1}{T_1} = \frac{V_2}{T_2}$$

If the volume is fixed the pressure of the gas will increase with increase of temperature because the molecules, moving faster, will hit the walls of the containing vessel harder and more often.

Pressure law

All gases increase in pressure by 1/273 of their pressure at 0°C for each degree rise in temperature, provided the volume is kept constant.

Similarly, extending this to lower temperatures, a gas will have zero pressure at −273°C.

The pressure law can be restated in terms of the Kelvin scale of temperature: at constant volume, the pressure of a gas is directly proportional to the Kelvin temperature;

or: $\frac{p_1}{T_1} = \frac{p_2}{T_2}$

We can picture the molecules of a gas at absolute zero as not having any kinetic energy, therefore not moving. In practice, all gases liquefy before this temperature is reached.

General gas equation

If, as usually happens in practice, pressure, volume and temperature all vary, the general gas equation should be used.

$$\frac{p_1 \times V_1}{T_1} = \frac{p_2 \times V_2}{T_2}$$

Kelvin temperature must be used for this equation.

Transmission of heat

Conduction usually takes place in solids. Metals are good conductors. Non-metals are poor conductors, or insulators.

Convection can take place only in liquids and gases. Owing to the increase in volume when heated, fluid becomes less dense and rises. The reason for land and sea breezes and the action of hot-water systems can be explained by convection.

Radiation All electromagnetic waves carry energy. Heat energy from the sun or any other hot body is carried by electromagnetic waves of a particular wavelength, known as **infra-red**. These have the same properties as light: they travel at the same high speed and can be reflected and refracted. They have a wavelength a little longer than visible red light.

Dull, dark surfaces absorb light very well. They also absorb radiated heat. On the other hand, shiny, polished and white surfaces that reflect light also reflect the radiated heat.

Dull surfaces will emit radiation well when hot. A polished surface at the same temperature will not emit radiation so well.

Heat energy

Joule showed that measured amounts of heat energy are produced by measured amounts of other forms of energy, making the old unit of heat energy, the kilocalorie, unnecessary. The joule can be used for heat energy as well as for other forms of energy.

The historical experiments of Joule, particularly that based on his paddle-wheel apparatus, and the definition of the kilocalorie are included on many examination syllabuses.

The **kilocalorie** is the amount of energy needed to heat 1 kg of water through 1°C.

Joule showed, though he used old British units, not joules and kilocalories, that 4 200 joules are the same as 1 kilocalorie.

The four-stroke internal combustion engine

Note Not all syllabuses include this topic, so check before learning it.

The complete cycle consists of the induction stroke, compression stroke, spark, power stroke and exhaust stroke. The chemical reaction between the petrol and air produces heat energy, pushing the piston down to produce rotational kinetic energy.

Specific heat capacity

When heat energy is supplied to a body, the temperature of the body increases. The supplied energy causes the molecules of the body to vibrate more rapidly. The actual rise in temperature which a given supply of heat energy causes will depend upon the number of molecules and the type of molecules present in the body.

The mass and specific heat capacity respectively are measures of the number and type of molecules present.

Specific heat capacity is the amount of energy needed to raise the temperature of 1 kilogram of the substance by 1°C. Units are joules/kg°C.

Therefore the specific heat capacity of water is 4 200 J/kg°C.

Heat energy = mass × specific heat capacity × temperature change.

Note that this energy is going into the substance to increase the random motion of the molecules, the internal energy.

Latent heat

When a kettle of water is boiling, but still switched on, the heat energy being supplied is no longer causing a rise in temperature, since this remains steady at the boiling point. Instead the energy is used to **change the state** of the water from liquid to vapour. The molecules of a liquid are held together by mutual forces of attraction, and to convert the liquid to a vapour energy must be supplied in order to do work in overcoming these forces. The molecules in a vapour are then much more free to move and spread out.

The energy required to convert a liquid into a vapour is called the **latent heat of vaporisation**.

The forces of attraction between the molecules of a solid are generally greater than those between the molecules of a liquid, so energy also has to be supplied to change a solid into a liquid. This energy is called the **latent heat of fusion**. The formal definitions of latent heats exclude changes in temperature. Thus the **specific latent heat of vaporisation** is defined as the **energy required to change the state of unit mass of a liquid to vapour without a change in temperature**, and the specific latent heat of fusion is defined in the same way, but for the change from a solid to a liquid. Since the definitions refer to energy change per unit mass, the units of specific latent heat are **joules/kilogram**.

Cooling produced by evaporation
As a liquid evaporates it takes its latent heat energy from its own internal energy, or the energy of its surroundings, thus causing a decrease in temperature. Molecules in a liquid have a range of speeds. The average speed determines the temperature. If the faster-moving molecules escape into the air, the slower-moving ones will remain in the liquid. The average speed, and hence the temperature, will therefore be lowered.

Saturated vapour pressure
At all temperatures the space above a liquid will contain its vapour. If the space is not limited the liquid will continue to evaporate and the vapour will diffuse away into the surrounding air. The molecules of vapour may hit the walls of a room thereby causing a pressure in addition to the air pressure. If however the space above the liquid is closed, a state of dynamic equilibrium is soon reached. Equal numbers of molecules are escaping and re-entering the liquid. The vapour is said to be saturated: it cannot hold any more molecules. The pressure a saturated vapour produces is called the **saturated vapour pressure**.

If the temperature is increased, the s.v.p. will also increase, for two reasons: there will be more molecules escaping into the vapour state and they will all be moving faster. Thus s.v.p. increases rapidly with temperature.

S.v.p. and boiling
A liquid will boil when its s.v.p. is equal to the pressure of the air above the liquid. The molecules have enough energy to form bubbles of vapour in the liquid.

Therefore the s.v.p. of water at 100°C is 760 mm.Hg.

Wave motion

The **frequency of a wave (f)** is the number of complete oscillations made in 1 second. The units are hertz (Hz).

The **wavelength** (λ) is the distance between two successive points on the wave which are in phase. The units are those of length, metres (m).

The **velocity of a wave** (v) is the velocity with which the wave moves forward.

The above three quantities are related by the formula $v = f\lambda$. The **amplitude of a wave** is its maximum displacement. The amplitude is a measure of the amount of energy carried by the wave.

Note In answering an examination question in which a drawing of wavefronts is required, never draw the wavefronts free-hand. Plane wavefronts should be drawn with a rule, circular wavefronts with compasses. Arrows should be drawn at right angles to the wavefronts to indicate the direction in which they are moving. (See pages 81–5 for the relevant diagrams.)

Refraction is the change in direction of a wave when it changes velocity. There is a corresponding change in wavelength. Remember that the frequency does not change. (See figure 44.)

Diffraction is the effect of waves bending round corners and spreading out after passing through narrow gaps. The longer the wavelength the more noticeable the effect.

Interference of two waves occurs when two waves of the same frequency overlap. They will interfere constructively if they are in phase: crest will meet crest and trough will meet trough. If, however, a crest meets a trough, they will interfere destructively because they are out of phase.

Constructive interference occurs if the waves from the two sources arrive at a point **in phase**. Destructive interference occurs if the waves from the two sources arrive **out of phase**.

To obtain coherent sources of light, that is, light sources with the same frequency and in phase with each other, the light must start from a single source. (See Young's slits experiment, page 86.)

Sound waves do not have crests and troughs like water waves. The vibration of the air molecules creates areas of high pressure, called **compressions**, and areas of lower pressure, called **rarefractions**.

These are called **longitudinal progressive waves**: the particles vibrate parallel to the direction in which the wave travels.

In **transverse progressive waves** the particles vibrate at right angles to the direction of travel of the wave.

The velocity of sound is about 330 m/s.

The velocity of light and other electromagnetic waves is 3×10^8 m/s.

Musical notes have **pitch** which is determined by their frequency. They have **loudness** which is determined by their amplitude. They have **quality** which is related to the shape of the the wave. The quality is caused by **overtones or harmonics** which are superimposed on the basic frequency.

Resonance is the setting into vibration of a body at its natural frequency by small impulses at that same frequency. Air will vibrate in a tube at certain frequencies only. If a tuning fork of the correct frequency is sounded close to the tube, resonance is heard.

Standing waves are produced in vibrating strings and in air in pipes, owing to an interference effect. The wave travelling in one direction interferes with its echo travelling in the other direction.

Places of constructive interference, large movement, are called **antinodes**.

Places of destructive interference, with no movement, are called **nodes**.

Electromagnetic waves can be classified into groups in order of decreasing wavelength as follows: radio waves, infrared, visible, ultraviolet, X-rays and γ-rays.

Light

Straight-line propagation

Shadows and eclipses occur because light travels in straight lines in any one transparent medium.

Note Though light is a wave motion, which can spread out and bend round corners, all rays in diagrams must be drawn with a ruler. Pinhole cameras work on this principle (see page 98).

The inverse square law for the intensity (I) of light falling on a surface at a distance (d) from a point source is a direct consequence of this linear motion. If the distance from the source is doubled the light energy has to cover four times the area and is thus $\frac{1}{4}$ as strong.
$I \propto 1/d^2$

Laws of reflection state that (1) the incident ray, the reflected ray and the normal are all in the same plane; (2) the angle of incidence is equal to the angle of reflection.

The image formed by a plane mirror is the same size as the object, as far behind the mirror as the object is in front, the same size, upright, virtual and laterally inverted.

Refraction is the bending of light rays as they pass from one medium into another. It results from the change in velocity of the light.

Snell's law states that

$$\frac{\text{sine of the angle of incidence } (i)}{\text{sine of the angle of refraction } (r)} = \text{a constant.}$$

This constant is called the refractive index of the medium and is given the symbol n.

It has since been shown that

$$n = \frac{\text{velocity of light in air}}{\text{velocity of light in the medium}}$$

Apparent depth

It can be shown that $\dfrac{\text{real depth}}{\text{apparent depth}} = n$

Note This is not the definition of refractive index.

Total internal reflection occurs when a ray of light in a medium is incident on the boundary between the medium and air at an angle greater than the **critical angle.**

At the critical angle the angle of refraction of the ray emerging into the air will be 90°.
$\therefore n = \sin 90°/\sin c$ ($\sin 90° = 1$)
$\mathbf{\sin c = 1/n}$
c for glass is 42° and for water 48°. (Remember that at angles less than the critical angle some internal reflection will occur, in addition to refraction.)

Reflection at curved mirrors
Note Not all examination syllabuses include this topic.

The focal length is half the radius of curvature.

Parallel rays reflect through or appear to come from the **focal point**.

For graphical solutions to problems, the paths of two rays should be traced. The point at which they cross shows the position of the corresponding point of the image.

One ray is the one which **passes through the centre of curvature**. It will hit the mirror normally (at right angles) and will thus be reflected back along its own path. **A second ray** is drawn **parallel to the principal axis**. After reflection it will pass through the focal point.

Remember that an image behind the mirror is **virtual**, while an image in front of the mirror is **real**.

Lenses
Note It is wise to check the relevant syllabus requirements, particularly with regard to calculations. Some boards require graphical solutions and some require formulae to be used.

A lens does not have a radius of curvature.

A lens has two focal points, one on either side. For a **convex** or **converging lens**, the focal points are **real**. Rays parallel to the principal axis pass through the focal point on the other side of the lens. For a **concave** or **diverging lens**, the focal points are **virtual**. Rays parallel to the principal axis appear to have come from the focal point after passing through the lens. Thus in the formula the value of f **for a convex lens is positive** and **for a concave lens is negative**.

For graphical solutions the paths of two rays should be traced from the top of the object. Where they cross gives the corresponding point of the image.

One ray is drawn **to pass through the pole**, or centre of the lens. It continues straight through, as it is undeviated by the lens.

A second ray is drawn **parallel to the principal axis**. After refraction it will, in the case of a convex lens, pass through the focal point on the other side of the lens. In the case of a concave lens it will appear to come from the focal point on the object side of the lens.

The lens formula is $\frac{1}{u}+\frac{1}{v}=\frac{1}{f}$. **Magnification** $= \frac{v}{u}$

Note Some syllabuses require candidates to know a method for finding the focal length of a lens.

Cameras have a convex lens which produces a small, real, inverted image on the light-sensitive film. It has an **iris diaphragm** which varies the size of the **aperture**. It has a **shutter** which opens briefly to allow light to fall on the film. The size of aperture and the shutter speed can both be varied to give the correct **exposure**. The lens produces a clear image of objects at a certain distance. For different object distances, the lens is moved closer to or further from the film.

The human eye is similar in some respects to the camera but different in others. It has a **lens** that produces the same type of image on a light-sensitive surface, the **retina**. The retina, however, is continually sensitive, sending pictures to the brain via the **optic nerve**. The size of the **pupil** is varied by the **iris**, depending on the intensity of the light. The lens-focusing mechanism is the main difference. To obtain clear images of objects at different distances, the **focal length** of the lens is changed by the **ciliary muscles** changing the curvature of the surfaces of the lens. This is called **accommodation**.

Note Questions on this topic occur frequently in O-level examination papers.

Defects of vision

Short sight Caused by the eye having too strong a lens. The eye can see things which are very close, but when looking at distant objects the image is formed **in front of the retina**.

To correct this, a **negative, concave lens** is put in front of the eye.

Long sight Caused by the eye having too weak a lens. It can produce a clear image of distant objects, but when looking at close objects the image is formed **behind the retina**. To correct this, an extra positive, convex lens is put in front of the eye.

Magnifying glass A single, strong convex lens. The object is placed **inside the focal point**, hence an enlarged, upright, virtual image is formed. This is optically the same as the eye-piece for both the microscope and the telescope.

Compound microscope This has two **strong, short-focal-length convex lenses**. The first lens, called the objective, produces a real enlarged image of the object. This image acts as the object for the eyepiece lens. The final image is virtual and inverted.

Astronomical telescope This has two convex lenses, a **weak, long-focal-length** lens and a **strong, short-focal-length** one. The weak objective lens produces a real image of the distant object at its focus. This is also the focal point of the eyepiece. The angular magnification is f_o/f_e. This final image is inverted. An inverter lens can be added or two prisms can be used to re-invert the image, as in prism binoculars.

Colour

The seven colours which make up the visible spectrum are red, orange, yellow, green, blue, indigo and violet. Red light has the longest wavelength of the seven, and is refracted least by a prism. Violet light has the shortest wavelength and is refracted most by a prism.

Red, green and **blue** are the three **primary colours**. If these three colours are added together, by shining a red lamp, a green lamp and a blue lamp together so that the patches of colour they produce overlap on a white screen, the region of overlapping appears white. If only two of the lamps are used: red + green = yellow; red + blue = magenta; blue + green = cyan. **Yellow, magenta** and **cyan** are the three **secondary colours**. Yellow = white − blue; magenta = white − green, cyan = white − red.

A pure red filter allows only red light to pass through it, and absorbs all other colours. If white light falls on a pure red surface, only the red component of the incident light is reflected, and the rest of the colours are absorbed.

Permanent magnetism

Permanent magnets, when freely suspended, will lie along the earth's magnetic field. The end pointing towards the north is called the north pole of the magnet, the one towards the south is called the south pole.

Like poles repel each other.
Unlike poles attract each other.

Magnetic materials are iron, steel, nickel, cobalt and certain iron salts.

Lines of force are directed **away from a north pole** and **towards a south pole**. They show the direction in which a free north pole would move if placed in a **magnetic field**.

Wilhelm Weber first suggested a molecular theory of magnetism. Each particle is regarded as being a small magnet. In a non-magnetic state, the small magnets form **closed chains**. When the substance is magnetised the magnets all point in one direction. This explains magnetic saturation and why poles appear at the centre when a magnet is broken in half.

In **magnetically soft iron** these little magnets are easily brought into line by an external field. But as soon as the external field is removed they will move out of line.

The reverse is true of **steel**. It is more difficult to line up the magnets, but once in line they will remain there, forming a permanent magnet.

The **earth** has a magnetic field around it. This magnetic field is similar in shape to that around a permanent magnet with its south pole close to the earth's north pole and north pole close to the earth's south pole.

A compass will point about 10° west of north in England. This is the **angle of declination**.

A compass needle suspended so that it can pivot vertically as well as horizontally will point down at an angle of 70° to the horizontal (in the UK). This is called the **angle of dip**.

A magnetic field has magnitude and direction. It is a **vector quantity**. Two fields added together produce zero at a **neutral** point.

Electrostatics

An **insulator** will not allow electricity to flow through it easily. Its electrons are held fairly firmly in place, although some can be added to or removed from its surface by **friction**.

A **conductor** will allow electricity to pass through it. Electrons in it are fairly free to move.

Two bodies with like charges will repel each other. Two bodies with unlike charges will attract each other.

Electrons will move in a conductor from regions of low potential to regions of high potential until it is all at the same potential.

When an insulator becomes electrified, it is said to be **charged**. Charge is measured in **coulombs**. If the insulator has an excess of electrons it is said to be negatively charged, and if it has a deficit of electrons it is said to be positively charged. When bodies are charged, they acquire a **potential**. Potential is measured in **volts**. A negatively charged body has a negative potential, and a positively charged body has a positive potential. Work has to be done in moving a positive charge towards a body with positive potential. Since work is measured in joules, there is a relationship between volts, joules and coulombs in that a volt is equivalent to a joule per coulomb, i.e.
volts = joules/coulombs.

The **capacitance** of a charged body is defined as the ratio:

$$\textbf{capacitance} = \frac{\textbf{charge}}{\textbf{potential}}.$$

Capacitance is measured in farads.

A gold-leaf electroscope measures difference in potential between the leaf and the earthed case.

Current electricity

If two points are maintained at different potentials, a **current** will flow if there is a path available. Current is measured in **ampères**, and one ampère corresponds to a current of 1 coulomb per second. The current, in amps, is related to the **potential difference** between the points, in **volts**, by **Ohm's law**. This states that the ratio of the potential difference across the ends of a conductor to the current in it is a constant provided that the temperature of the conductor does not alter. The **resistance** of a conductor is defined as the ratio:

$$\frac{\text{potential difference across its ends}}{\text{current flowing through conductor}}.$$

Resistance is measured in **ohms** (Ω).

For resistance in series $R = R_1 + R_2 + R_3 \ldots$ etc.

For resistances in parallel $\frac{1}{R} = \frac{1}{R_1} + \frac{1}{R_2} + \frac{1}{R_3} \ldots$ etc.

The **resistivity** of a material is defined as the resistance between opposite faces of a unit cube of the material. If ρ is the resistivity, A the cross-sectional area, R the resistance of the conductor and l its length, then

$$\rho = \frac{RA}{l}$$

Ammeters and voltmeters

An **ammeter** is a **low-resistance meter** which is used to measure **current**. It is always placed **in series** with the device through which the current is to be measured.

A voltmeter is a **high-resistance meter** which is used to measure **potential difference**. It is always placed **in parallel** with the device across which the p.d. is to be measured.

The **range** of an ammeter may be extended by placing a low-resistance **shunt** in **parallel** with the meter.

Electromotive force and internal resistance of a cell

A cell or accumulator is capable, by its chemical action, of producing a potential difference of a certain number of volts. This value is its **electromotive force**, or **e.m.f.** However, when the cell is in use, not all these volts are available to do useful work, since some are needed to drive the current through the cell itself. This is necessary since the cell has an **internal resistance**. The **volt** is defined in terms of the **work done in moving a unit charge**; thus the e.m.f. may be defined as the **work done in driving unit charge round the complete circuit**.

The work done in driving unit charge round the external circuit is equal to the potential difference across the external circuit, so: e.m.f. = internal p.d. + external p.d.

The internal p.d. is sometimes referred to as the 'lost volts'. When no current flows, there is zero internal p.d. since no work is required to drive current through the cell. This is why the simple definition of e.m.f.—as the potential difference on open circuit—is sometimes used.

The **potentiometer** is used to measure e.m.f. Adjustment is made until no current flows, so a potentiometer may be used to compare e.m.f.s.

The internal resistance of a cell may be defined as the resistance of the chemicals within the cell. It is measured in ohms, and is usually fairly small.

Suppose a cell of e.m.f. E volts and internal resistance r ohms is connected to an external circuit of resistance R ohms and a current of I amps flows.

From the definition of resistance, i.e. potential difference/current, it follows that the external p.d. is IR volts and the internal p.d. is Ir volts.
Hence $E = Ir + IR = I(r + R)$

When an electromotive force exists, current will flow only if there is a complete circuit. This is why, when describing the action of such electrical devices as transformers and dynamos, it is important to refer to **induced e.m.f.s** and not to induced currents.

Heating effect of a current

Electricity is a flow of electrons from a place of low potential to one of higher potential. When electrons flow through a resistor, the number of electrons which leave the resistor is the same as the number which entered, and ammeters placed at either end of the resistor would both give the same reading. But since work is done in driving the electrons through the resistor, the electrons which leave have less energy than those which enter. The energy which is removed from the electrons is converted into heat, light or some other form by the resistor. There is therefore a **potential drop** across the ends of the resistor.

The energy in joules = potential difference in volts × charge in coulombs.

Since charge (in coulombs) = current (in amps) × time (in seconds)
energy = potential difference × current × time.
Joules = volts × amps × seconds.

Power is the rate of doing work, or the rate of using energy.

Power = energy/time, so
power = potential difference × current.
Watts = volts × amps.

In practical terms, the joule and watt are both rather small units. It is quite usual therefore to find kilowatts and kilojoules being used, and for some purposes megawatts and megajoules.

An alternative unit for energy is the kilowatt.hour (kWh).

Since 1 000 watts = 1 kW, and 3 600 seconds = 1 hour, it follows that 1 kWh = 3 600 000 joules.

Chemical effect of a current
Electricity in liquids and gases is carried by **ions**. An ion is an atom, or group of atoms, which has gained or lost one or more electrons. It is the outer, or valence, electrons which may be lost, and the number which is gained or lost depends on the **valency** of the atom or group. A common ionic substance is copper sulphate, $CuSO_4$. Copper sulphate crystals consist of doubly charged copper ions, Cu^{++}, and sulphate ions, SO_4^{--}. It is the strong electrostatic force of attraction between the oppositely charged ions which makes copper sulphate very stable. When the crystals are dissolved in water, the ions still exist in the solution. If an electric current is passed through a solution using copper electrodes, the positively charged copper ions migrate to the negatively charged electrode (**cathode**), and the negatively charged sulphate ions migrate to the positively charged electrode (**anode**). Copper is deposited on the cathode and removed from the anode.

Copper sulphate solution is an **electrolyte**. This term may be applied to any substance which passes a current and is decomposed by that current.

The **electrochemical equivalent**, e.c.e., of a substance is the mass deposited during electrolysis by one coulomb of charge. As we have already seen, charge (in coulombs) = current (in amps) × time (in seconds) so if M kg of a substance are deposited by a current of I amps flowing for t seconds, then $M = zIt$ where z is the electrochemical equivalent.

Faraday's laws of electrolysis state that:
(1) the mass of a substance deposited in electrolysis is proportional to the quantity of charge flowing in the circuit (it is this law which provides the definition of e.c.e.);

(2) the masses of different substances deposited by the same quantity of charge are proportional to their chemical equivalents.

Magnetic effects of currents

The magnetic field resulting from a current passing through a long straight wire is circular around the wire. The direction of the field is given by the **corkscrew rule**.

The magnetic field resulting from a current flowing through a **solenoid** is similar to that of a bar magnet on the outside of the coil with a strong uniform field inside the coil.

Motor effect

A conductor carrying an electric current in a magnetic field will have a force acting on it. The direction of the force is stated by **Fleming's left-hand rule**. (**Note** Do not confuse this with the right-hand rule. It is advisable to state which rule you are using so that even if you show the force going in the wrong direction you will get marks for knowing which rule to use.)

A **D.C.** motor uses a **split-ring commutator** and **carbon brushes** and the coils are wound on a **laminated soft iron core**. (The examiner will look for correct use of the terms in bold print; he will also check that the split of the ring is shown in the correct position to give a reversal of the current when the coil is at 90° to the magnetic field.)

Electromagnetic induction

A conductor moving relative to a magnetic field will have an **induced e.m.f.** across its ends. If the two ends are joined outside the magnetic field a current will flow. The motion, magnetic field and current are all at 90° to each other.

Fleming's right-hand rule states the directions of the motion, the magnetic field and the induced e.m.f.

Faraday's law states the magnitude of the induced e.m.f. The main thing to remember is that there is an induced e.m.f. only when a change in the magnetic field is taking place, or the conductor is moving through the magnetic field at an angle to it.

For working out what is happening **Lenz's law** is useful. If electrical energy is being produced, mechanical energy must be used up. Lenz's law can also be used to predict the direction of the induced e.m.f.

Note The principles governing A.C. and D.C. generators are often the subject of examination questions. Check that you are sure why continuous-ring commutators are used for A.C. and split-ring commutators for D.C.

Transformers

Remember the equation where V_1 and V_2 are the primary and secondary voltages, and N_1 and N_2 are the primary turns and secondary turns.

$$\frac{V_1}{V_2} = \frac{N_1}{N_2}$$

If N_2 is greater than N_1 the voltage is increased: the instrument will then be called a step-up transformer. But remember that the current is correspondingly decreased. $V \times I$ is the power and that cannot increase.

Note Transformers are another common examination topic. Candidates are likely to be asked whether A.C..or D.C. is used and why, and also why laminated iron cores are used.

Power transmission

For the same power as that quoted above, $V \times I$, if voltage is very high current will be very low. It is the current in the transmission lines that causes heating, with the resulting loss of electrical energy. Thus the higher the voltage the smaller the current, and the more efficient the system.

Electronics

Millikan's experiment

By balancing the gravitational force on a charged oil drop by an upward electrostatic force, Millikan calculated the charge on a large number of such drops. The results showed that, in each case, the total charge on the drop was a multiple of a particular value. He deduced that this was the smallest quantity of charge that could exist, that is, the charge carried by a single electron. He found its value to be $-1{\cdot}6 \times 10^{-19}$ coulombs.

Cathode-ray tube

The following features might be the scoring points in an examination question about this instrument and should be shown on any diagram pertaining to it.

(1) The **electron gun** including the **heater**, **cathode**, and **anode**. The diagram should include the information, or mention should be made of the fact that the **anode is maintained at a high positive potential** with respect to the cathode.

(2) The **deflection system**. The X-plates or a magnetic field deflects the beam horizontally, and the Y-plates deflect it vertically.
(3) The **fluorescent screen**. Fluorescent material is used to coat the inside of the screen.
(4) The **highly evacuated tube**.

Diode valve

The essential features of a diode valve are (i) a heater, (ii) a cathode, (iii) an anode. These are contained in an evacuated tube.

When the heater is in operation, electrons are thermionically emitted from the cathode, and form a **space charge**. Since these electrons are negatively charged, they will be attracted by the anode if the latter is maintained at a positive potential with respect to the cathode, so a current will flow across the valve. But if the anode is negatively charged, it will not attract the electrons so no current will flow. This is the basis of the use of a diode valve as a **rectifier**, i.e. to change alternating current into direct current.

Radioactivity

Atoms consist of a central **nucleus**. This contains positively charged **protons**, and except for hydrogen atoms, **neutrons** which have no charge. Negatively charged electrons orbit the nucleus, and the number of electrons is equal to the number of protons, under normal circumstances. This number is called the **atomic number** of the atom. Thus the atomic number of hydrogen is 1, that of helium is 2, and so on. The total number of protons and neutrons together is called the **mass number**. Hydrogen, having one proton and no neutrons, has a mass number of 1. Helium, having two protons and two neutrons, has a mass number of 4.

It is the atomic number which gives an atom its chemical characteristics. The atomic number of carbon is 6, and any atom with 6 protons in its nucleus is chemically identifiable as carbon. Normal carbon atoms have 6 neutrons as well, giving these atoms a mass number of $6+6=12$. But some carbon atoms have two extra neutrons in their nuclei. This does not affect their chemical properties, but does alter their physical properties, since the latter type have a mass number of 14. The two types of carbon atoms, carbon-12 and carbon-14, are said to be **isotopes** of carbon.

Radioactive isotopes are those which have unstable nuclei which break up. In the process α, β and γ radiation may be emitted.

α-particles are **helium nuclei** consisting of **two protons and two neutrons**. They carry a **positive** charge. They cause **heavy ionisation** in air, and for this reason have only a **short range**. They may be easily stopped, e.g. by a thin piece of paper. α-particles **may be deflected by electric or magnetic fields**, the direction of deflection indicating their positive charge.

β-particles are **negatively charged electrons**, whose mass is much less than that of α-particles, and so have a **greater range** in air. They are able to **pass through paper and thin metal foil**. β-particles may also be deflected **by electric and magnetic fields**, but in the opposite direction to α-particles, showing that they are negatively charged.

For the same field, β-particles are deflected more than α-particles since they are less massive.

γ-rays are **electromagnetic waves**, similar to light but of much **shorter wavelength**. They are **uncharged**, so **cannot be deflected by electric or magnetic fields**. They are the **most penetrating** of the three rays, and can pass through several centimetres of lead. They **produce ionisation**, though only to a very limited extent.

The **half-life** of an isotope may be defined as the time taken for the rate of decay of a sample of the material to fall to half its original value, or the time for half the radioactive isotopes in a given sample to decay away. It should not be defined in terms of mass, because the radioactive atoms do not decay away to nothing; most of their mass is retained in the sample.

Methods of detection

(1) **Photographic emulsions** are affected by radioactive emissions.
(2) **Gold-leaf and pulse electroscopes** discharge when the air becomes conducting because of the ionising effect of radiations.
(3) **Spark counter and Geiger-Müller tube** indicate the passing of individual particles, by using their ionising effect to produce an electrical discharge.
(4) **The cloud chamber** shows the actual paths of individual particles, by leaving a trail of condensed vapour droplets on the ions.

Other study aids in the  series

KEY FACTS CARDS

30p: Woodwork
Metalwork
Henry IV Part I
Henry V
Merchant of Venice
Richard II
Richard III
Twelfth Night
35p: Latin
German
Macbeth
Julius Caesar
40p: New Testament
45p: Geography – Regional
English Comprehension & Precis
English Language & Exam Essay
Algebra
45p: Economics
50p: Elementary Mathematics
Modern Mathematics
English History (1815–1939)
Chemistry
Physics
Biology
Geometry
Geography
French
Arithmetic & Trigonometry
General Science
Additional Mathematics
Technical Drawing

KEY FACTS COURSE COMPANIONS

40p: Economics
50p: Modern Mathematics
50p: Algebra
50p: Geometry
55p: Arithmetic & Trigonometry
Additional Mathematics
55p: Geography
French
Physics
Chemistry
English
Biology

KEY FACTS A-LEVEL BOOKS

55p: Chemistry
Biology
Pure Mathematics
Physics

All **KEY FACTS** titles are published by

ibp Intercontinental Book Productions
Berkshire House, Queen Street, Maidenhead, SL6 1NF
in conjunction with the distributors, Seymour Press Ltd., 334 Brixton Road, London, SW9 7AG

Prices are correct at time of going to press.